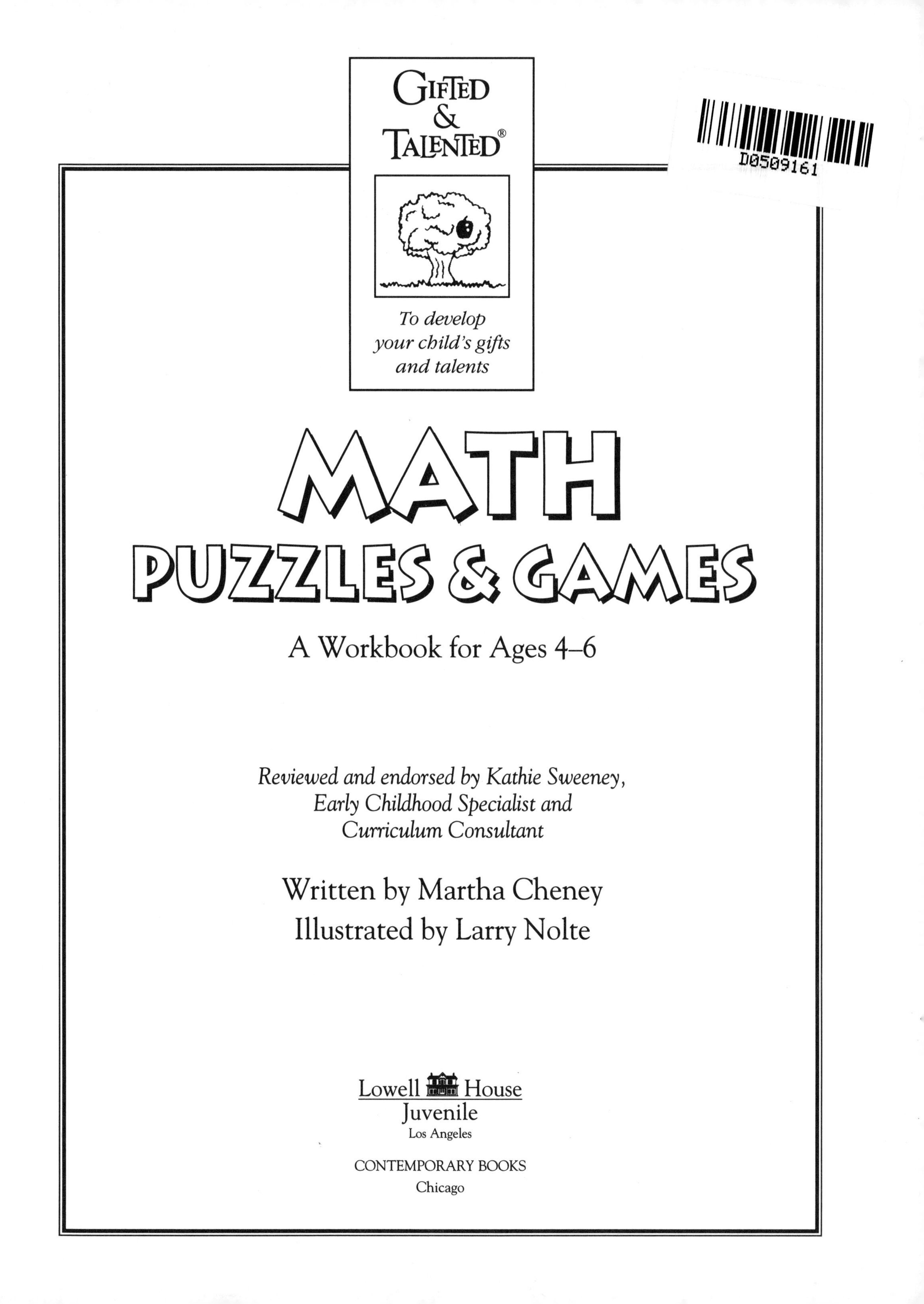

MATH PUZZLES & GAMES

A Workbook for Ages 4–6

*Reviewed and endorsed by Kathie Sweeney,
Early Childhood Specialist and
Curriculum Consultant*

Written by Martha Cheney
Illustrated by Larry Nolte

Lowell House
Juvenile
Los Angeles

CONTEMPORARY BOOKS
Chicago

Requests for such permissions should be addressed to:
Lowell House Juvenile
2020 Avenue of the Stars, Suite 300
Los Angeles, CA 90067

Lowell House books can be purchased at special discounts when ordered in bulk for premiums and special sales. Contact Department TC at the above address.

Manufactured in the United States of America

ISBN: 1-56565-500-1

10 9 8 7 6 5 4 3

Note to Parents

GIFTED & TALENTED® WORKBOOKS will help develop your child's natural talents and gifts by providing activities to enhance critical and creative thinking skills. These skills of logic and reasoning teach children **how** to think. They are precisely the skills emphasized by teachers of gifted and talented children.

Thinking skills are the skills needed to be able to learn anything at any time. Unlike events, words, and teaching methods, thinking skills never change. If a child has a grasp of how to think, school success and even success in life will become more assured. In addition, the child will become self-confident as he or she approaches new tasks with the ability to think them through and discover solutions.

GIFTED & TALENTED® WORKBOOKS present these skills in a unique way, combining the basic subject areas of reading, language arts, and math with thinking skills. The top of each page is labeled to indicate the specific thinking skill developed. Here are some of the skills you will find:

- Deduction—the ability to reach a logical conclusion by interpreting clues
- Understanding Relationships—the ability to recognize how objects, shapes, and words are similar or dissimilar; to classify or categorize
- Sequencing—the ability to organize events, numbers; to recognize patterns
- Inference—the ability to reach a logical conclusion from given or assumed evidence
- Creative Thinking—the ability to generate unique ideas; to compare and contrast the same elements in different situations; to present imaginative solutions to problems

How to Use GIFTED & TALENTED® WORKBOOKS

Each book contains activities that challenge children. The activities range from easier to more difficult. You may need to work with your child on many of the pages, especially with the child who is a non-reader. However, even a non-reader can master thinking skills, and the sooner your child learns how to think, the better. Read the directions to your child and, if necessary, explain them. Let your child choose to do the activities that interest him or her. When interest wanes, stop. A page or two at a time may be enough, as the child should have fun while learning.

It is important to remember that these activities are designed to teach your child **how to think**, not how to find the right answer. Teachers of gifted children are never surprised when a child discovers a new "right" answer. For example, a child may be asked to choose the object that doesn't belong in this group: a table, a chair, a book, a desk. The best answer is **book**, since all the others are furniture. But a child could respond that all of them belong because they all could be found in an office or a library. The best way to react to this type of response is to praise the child and gently point out that there is another answer, too. While creativity should be encouraged, your child must look for the best and most **suitable** answer.

GIFTED & TALENTED® WORKBOOKS have been written by teachers. Educationally sound and endorsed by leaders in the gifted field, this series will benefit any child who demonstrates curiosity, imagination, a sense of fun and wonder about the world, and a desire to learn. These books will open your child's mind to new experiences and help fulfill his or her true potential.

Look at this **cluster,** or group, of shapes.

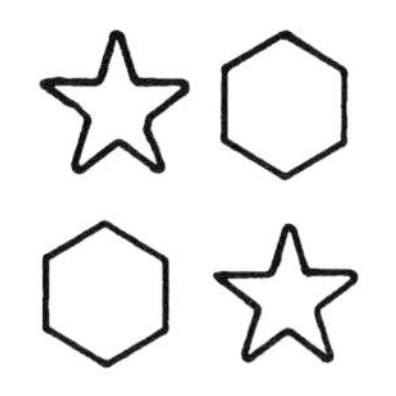

How many exact copies of this cluster can you find in the group of shapes below? An exact copy must contain the same shapes in the exact same design. Draw a circle around each matching cluster. One is done for you.

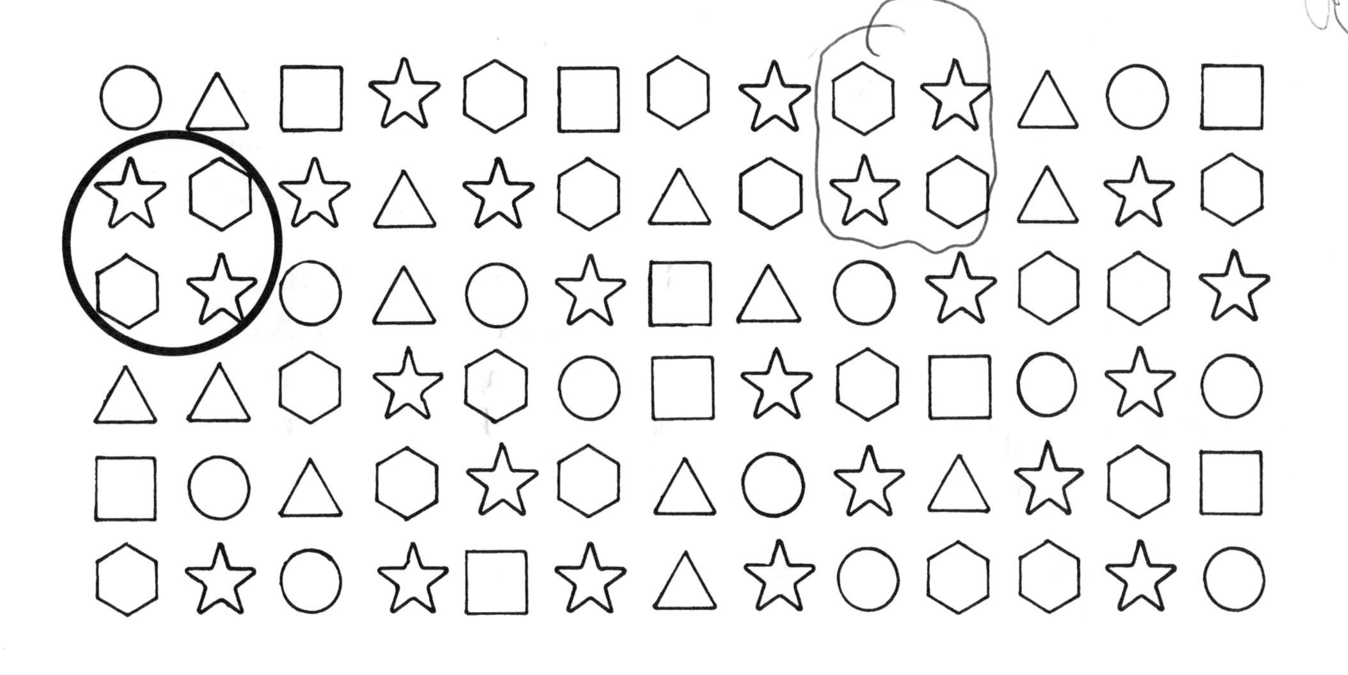

How many clusters are there? ___________

Look at this **cluster,** or group, of shapes.

How many exact copies of this cluster can you find in the group of shapes below? An exact copy must contain the same shapes in the exact same design. Draw a circle around each matching cluster. One is done for you.

How many clusters are there? 5

Look at this **cluster,** or group, of shapes.

◐ ⦶ ◑

How many exact copies of this cluster can you find in the group of shapes below? An exact copy must contain the same shapes in the exact same design. Draw a circle around each matching cluster. One is done for you.

How many clusters are there?___________

For an extra challenge, create your own cluster of shapes on a separate piece of paper. Then draw a pattern of shapes that contains at least two exact copies of your cluster. Ask a friend or family member to find and circle the matching clusters inside your pattern.

Analogies are puzzles that show the relationship between certain figures, words, or objects.

Here is an example:

Circle the figure that best completes each analogy below.

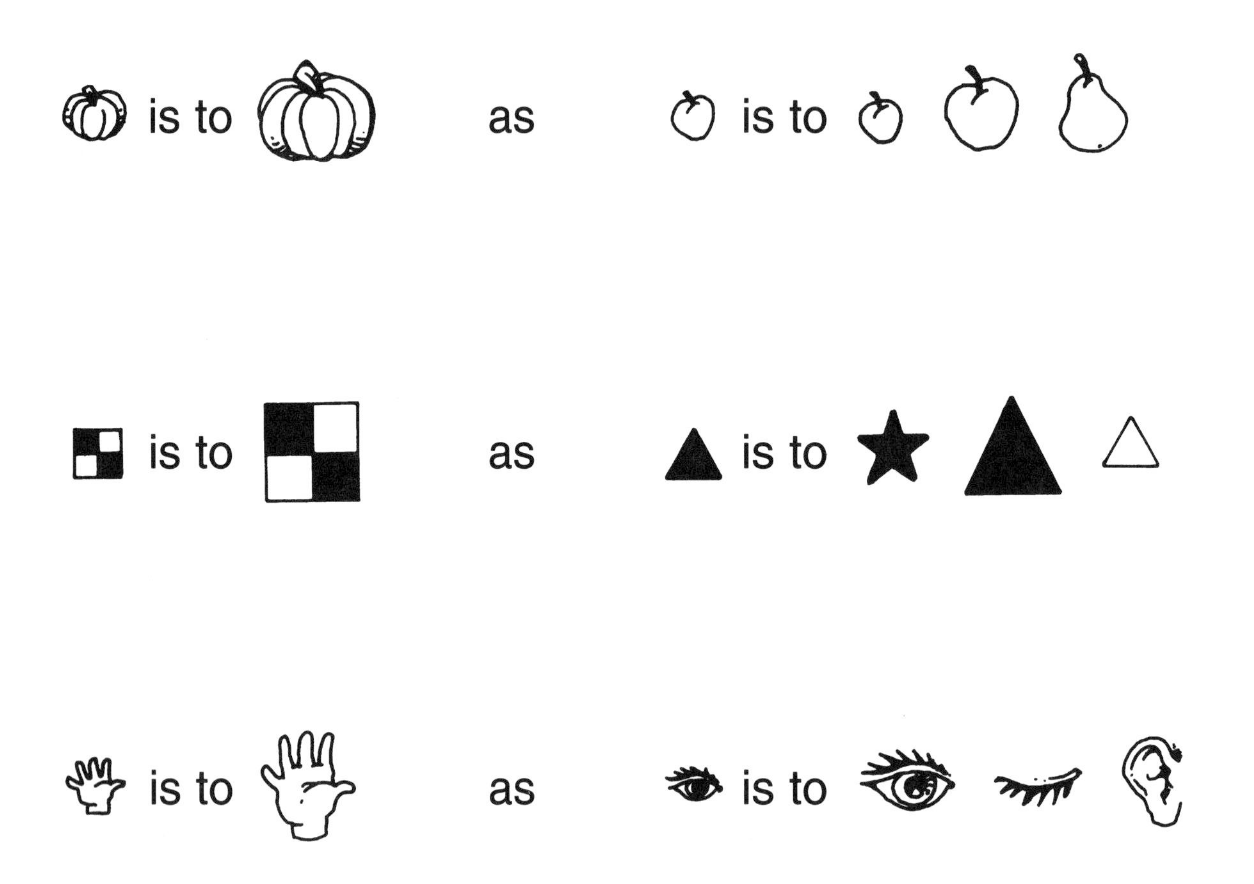

Analogies are puzzles that show the relationship between certain figures, words, or objects.

Here is an example:

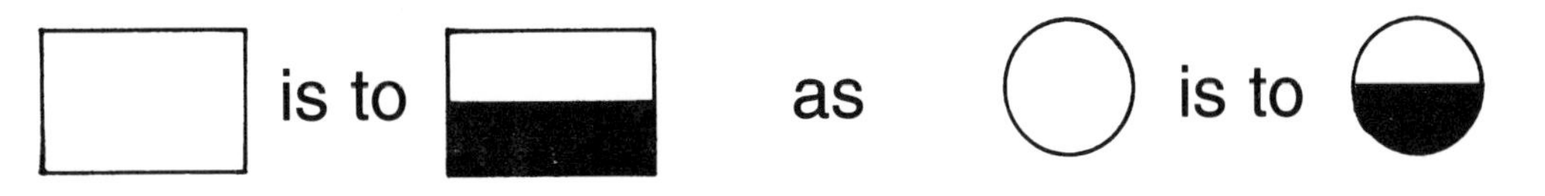

Circle the figure that best completes each analogy below.

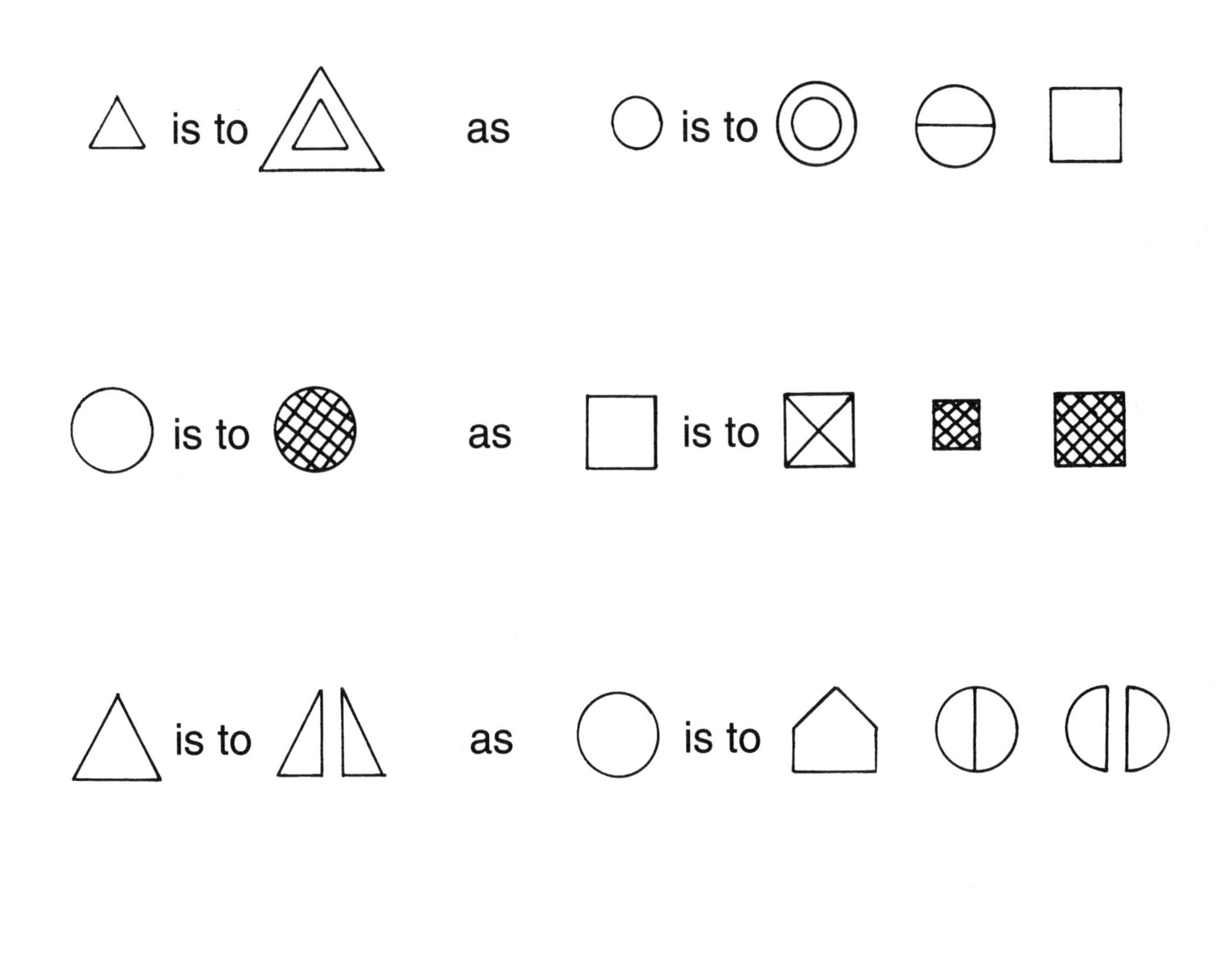

Analogies are puzzles that show the relationship between certain figures, words, or objects.

Here is an example:

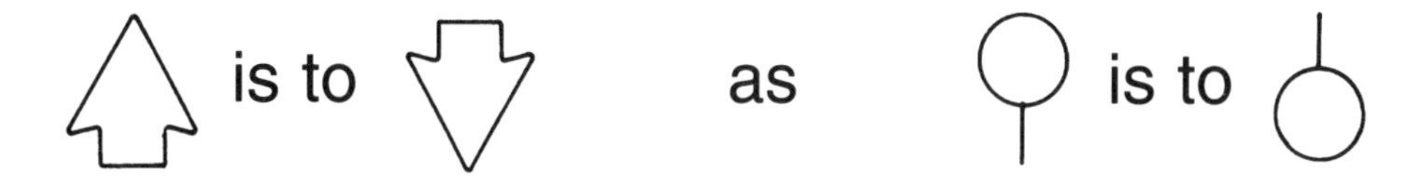

Circle the figure that best completes each analogy below.

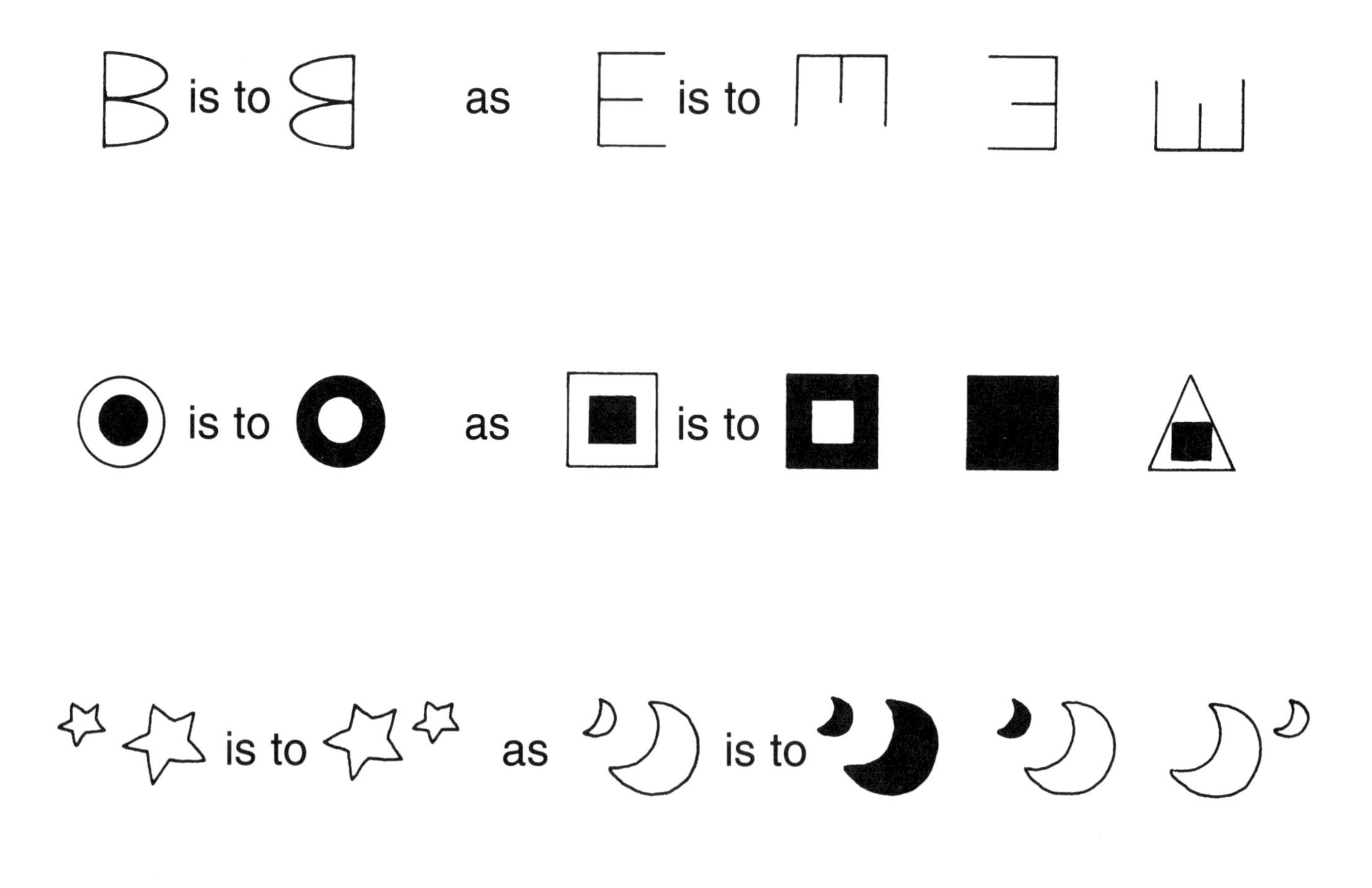

Analogies are puzzles that show the relationship between certain figures, words, or objects.

Here is an example:

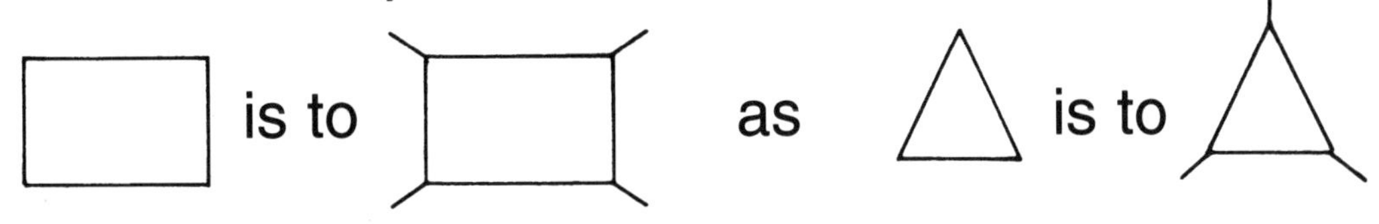

Circle the figure that best completes each analogy below.

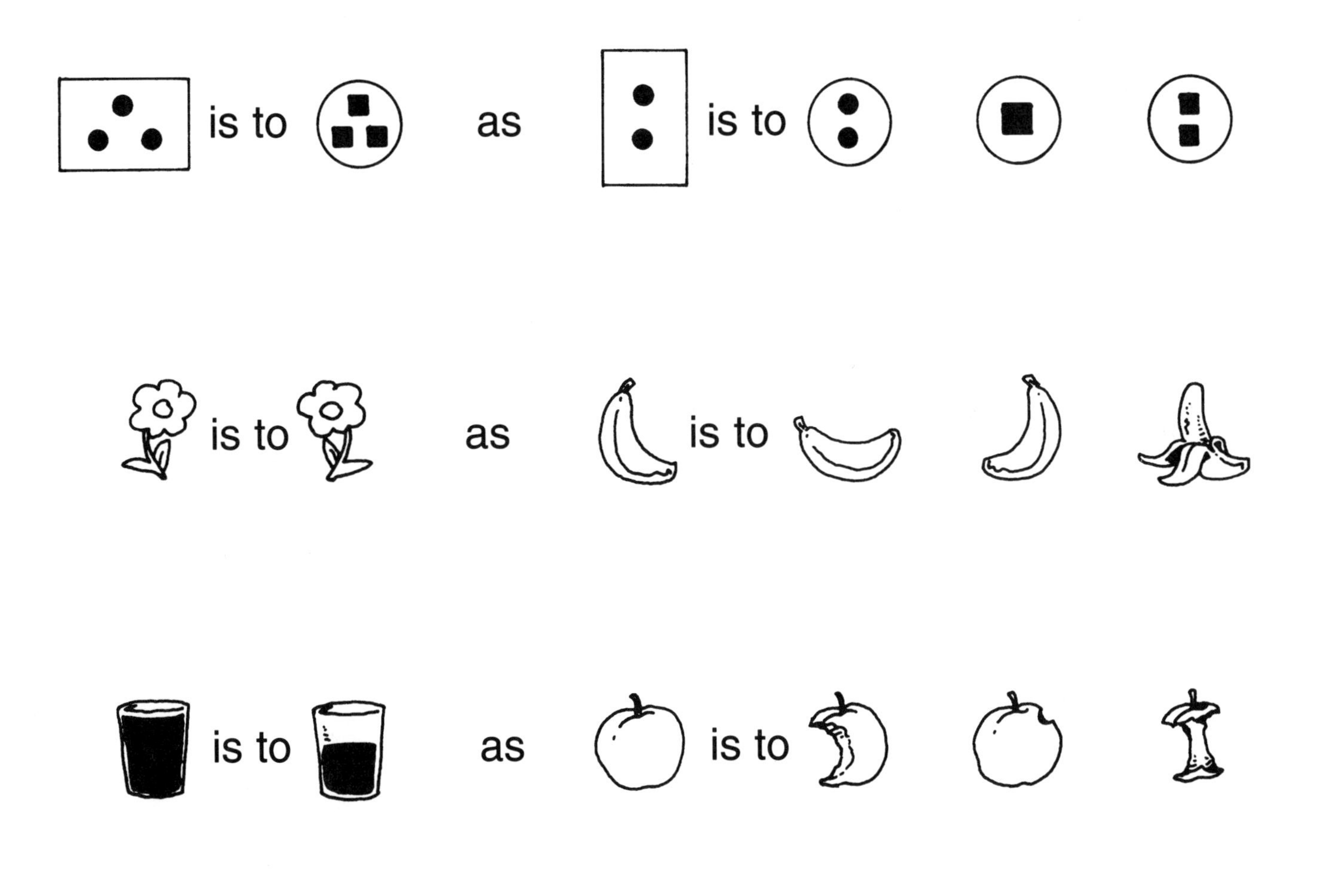

Mr. Hill always hangs his laundry out in the sun to dry. Color each line of laundry with crayons to create an interesting pattern. Use at least three different colors for each pattern. The row of socks has been started for you. Continue the pattern to complete the row, and then color each sock any way you like. Color the other rows of laundry using your own patterns.

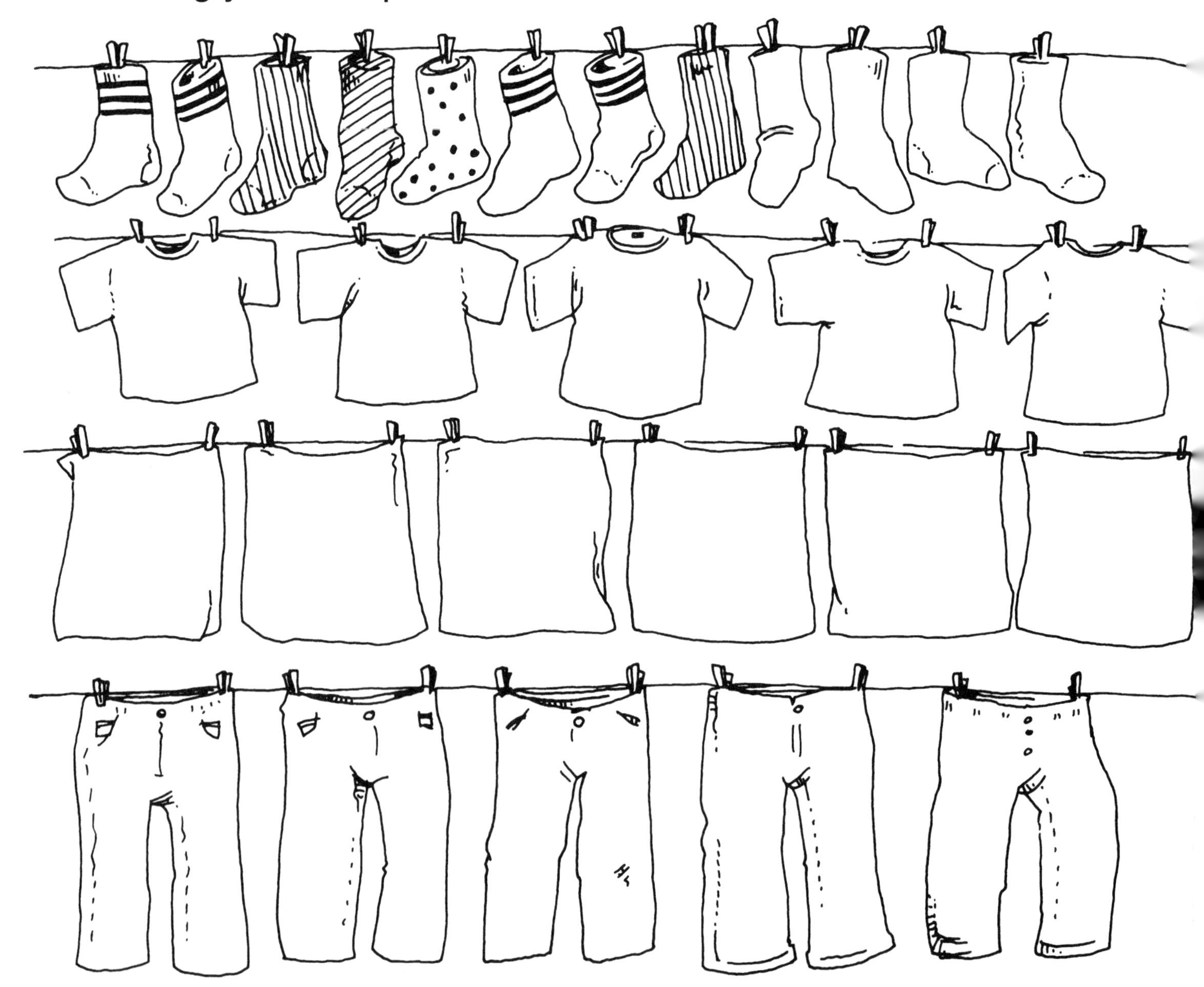

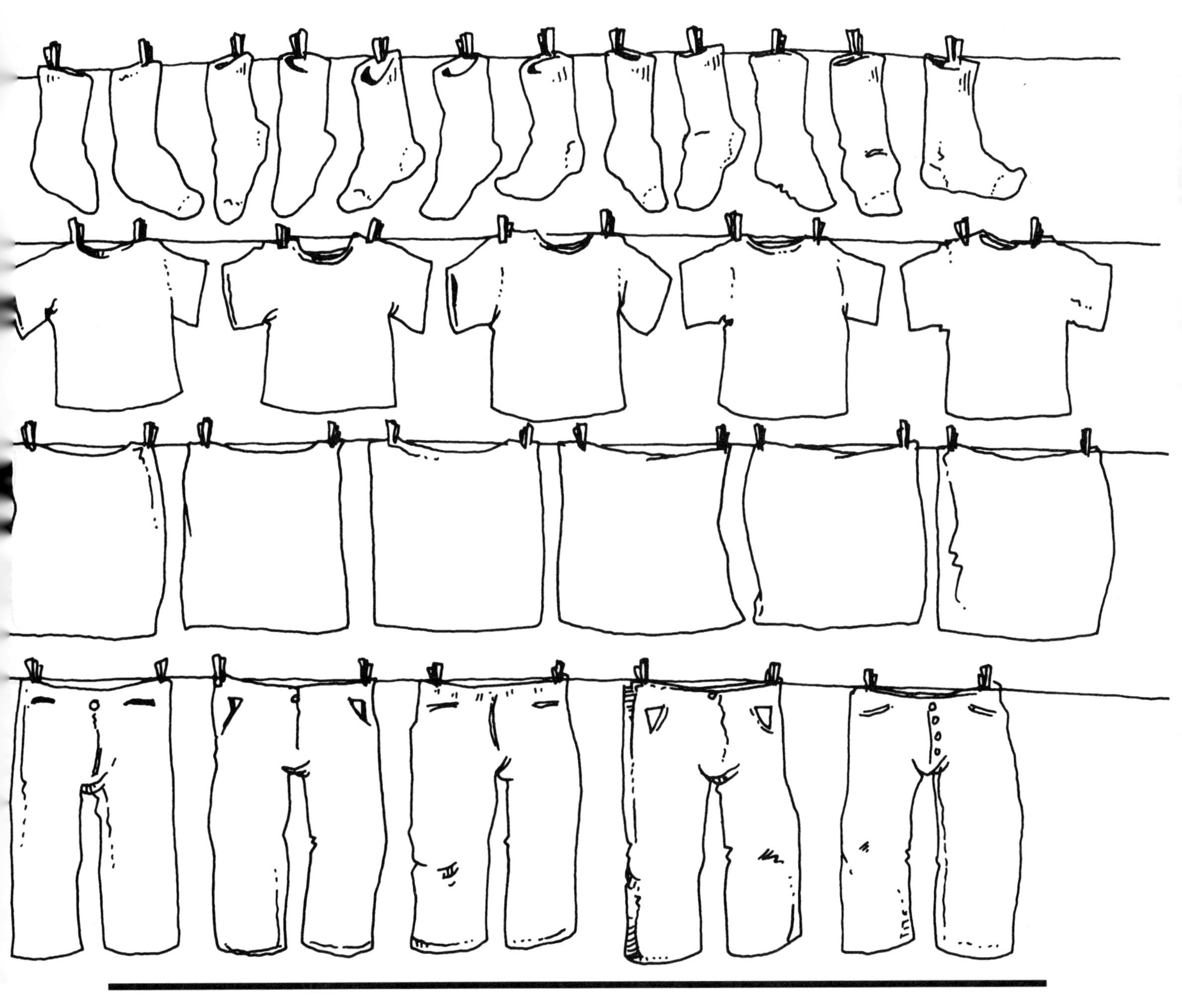

Pretend that you are blindfolded. The tray of chocolate and vanilla cookies below is set in front of you. You are allowed to take only **1** cookie. But since you are blindfolded, you cannot see which kind of cookie you take.

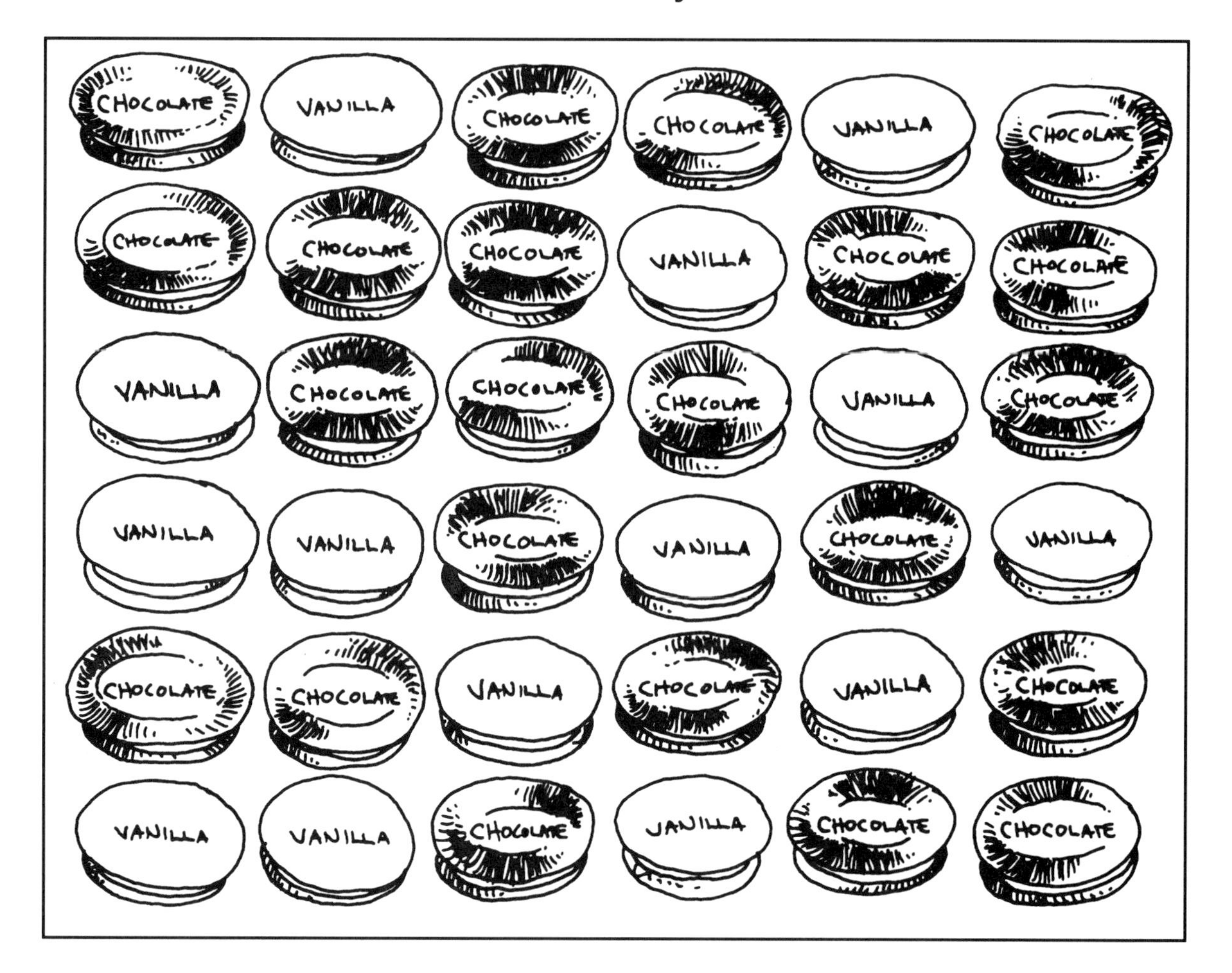

Which kind of cookie do you think you are most likely to pick? ______________________________

Why? ______________________________

____

In the figures below, triangles have been joined together to form squares. Use crayons to color each square blue and green. Color each square completely. Do not leave any white areas. Make sure that each square is colored differently from all the others.

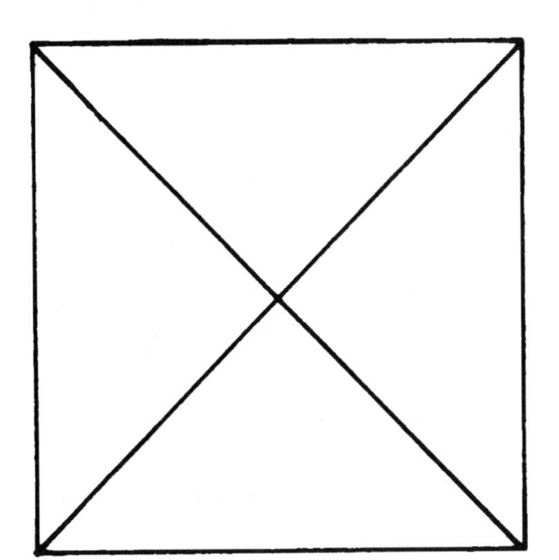

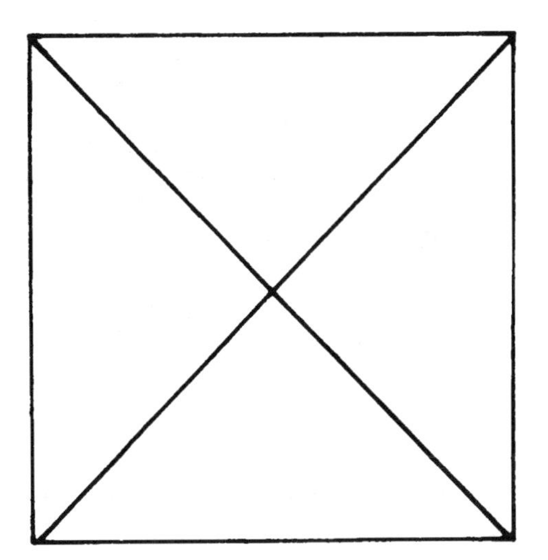

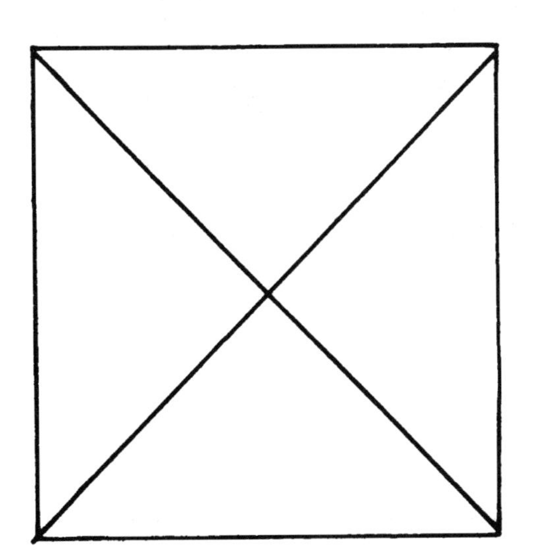

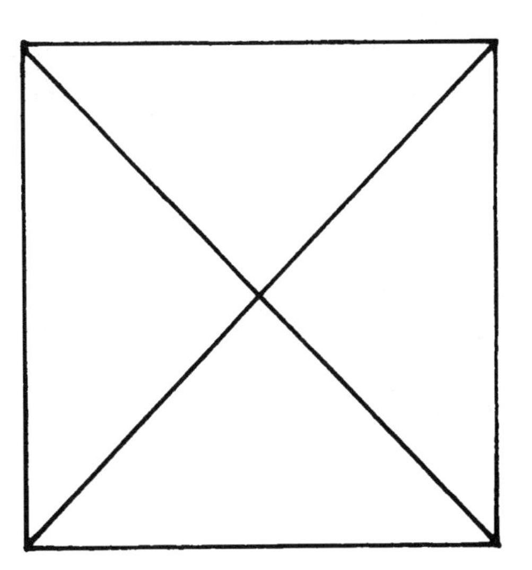

In the figures below, triangles have been joined together to form rectangles. Use crayons to color each rectangle orange and red. Color each rectangle completely. Do not leave any white areas. Make sure that each rectangle is colored differently from all the others.

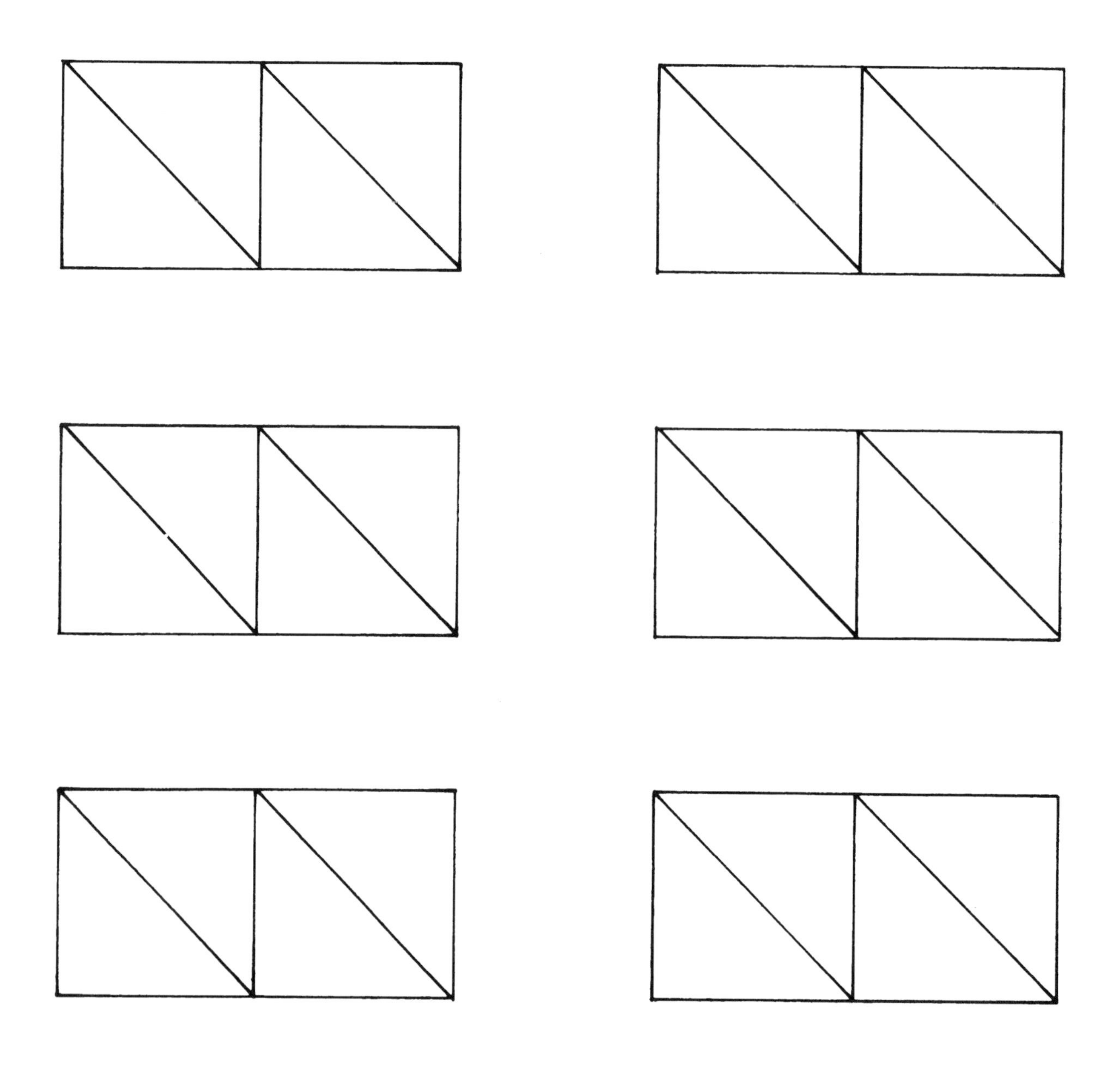

Here is another game to play with triangles. How many triangles shaped like this △ would it take to form each shape below? Write that number on the line beneath each shape. The first one is done for you.

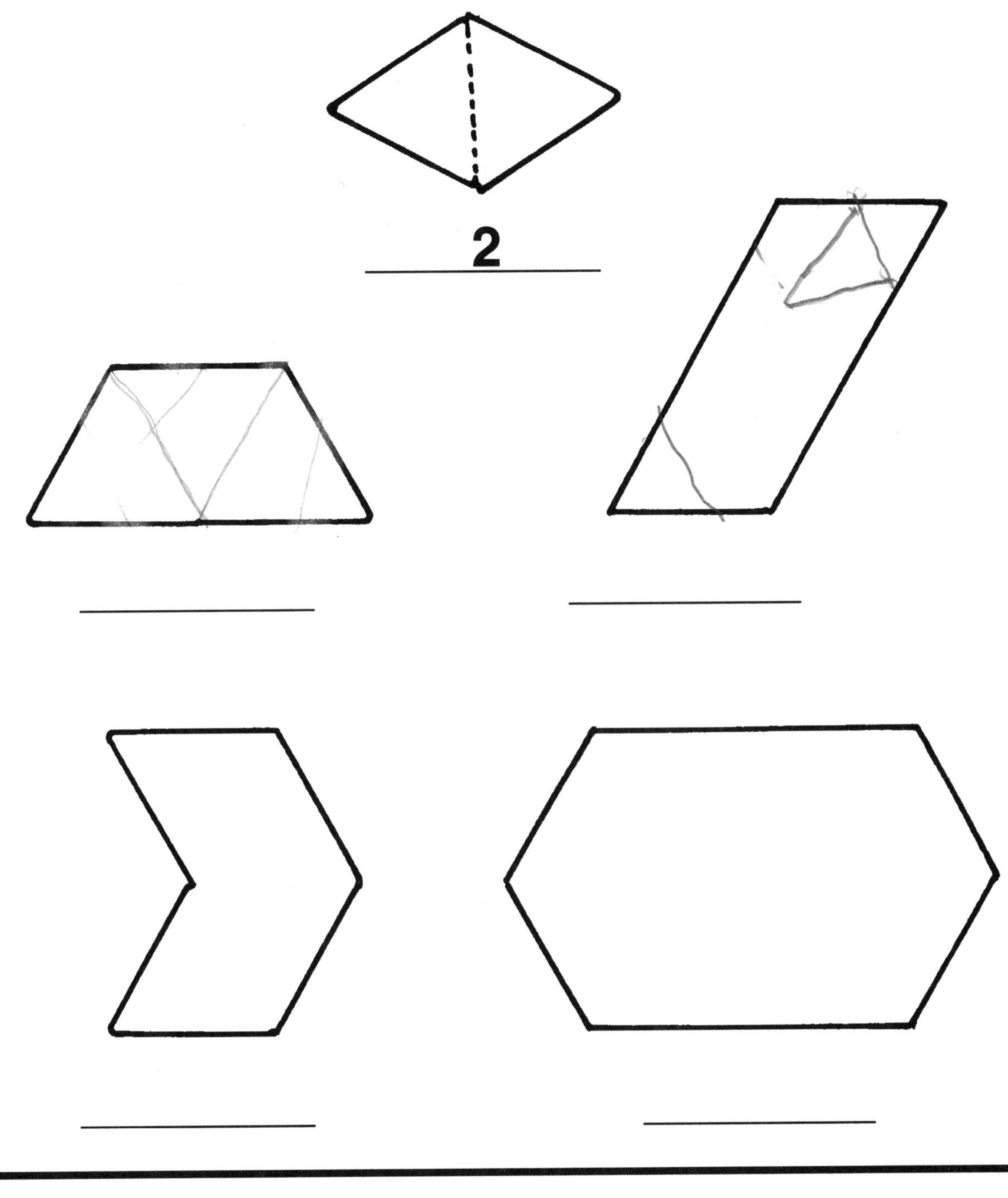

The house next door to Michael's house is similar to his own house in some ways and different from it in other ways.

This is Michael's house.

How are Michael's house and the house next door alike?

____

____

____

How do the houses differ? ______________________

____

____

____

Look at the objects on Keri's breakfast table. They are alike in some ways and different in other ways.

How are the objects alike? ______________________________

____

____

____

How do the objects differ? ______________________________

____

____

____

Josh and Heather each have an ice-cream cone. The ice-cream cones are alike in some ways and different in other ways.

How are Josh's cone and Heather's cone alike? ________

____

____

How do their ice-cream cones differ? ________________

____

____

For an extra challenge, on a separate piece of paper draw two pictures that are alike in some ways and different in other ways. Then ask a friend or family member to describe how they are the same and how they are different.

Look at the numerals on this page. They are all mixed up! Some of them are right-side up, some are upside down, and some are even sideways! How many of each numeral can you find? Write your answers in the spaces below.

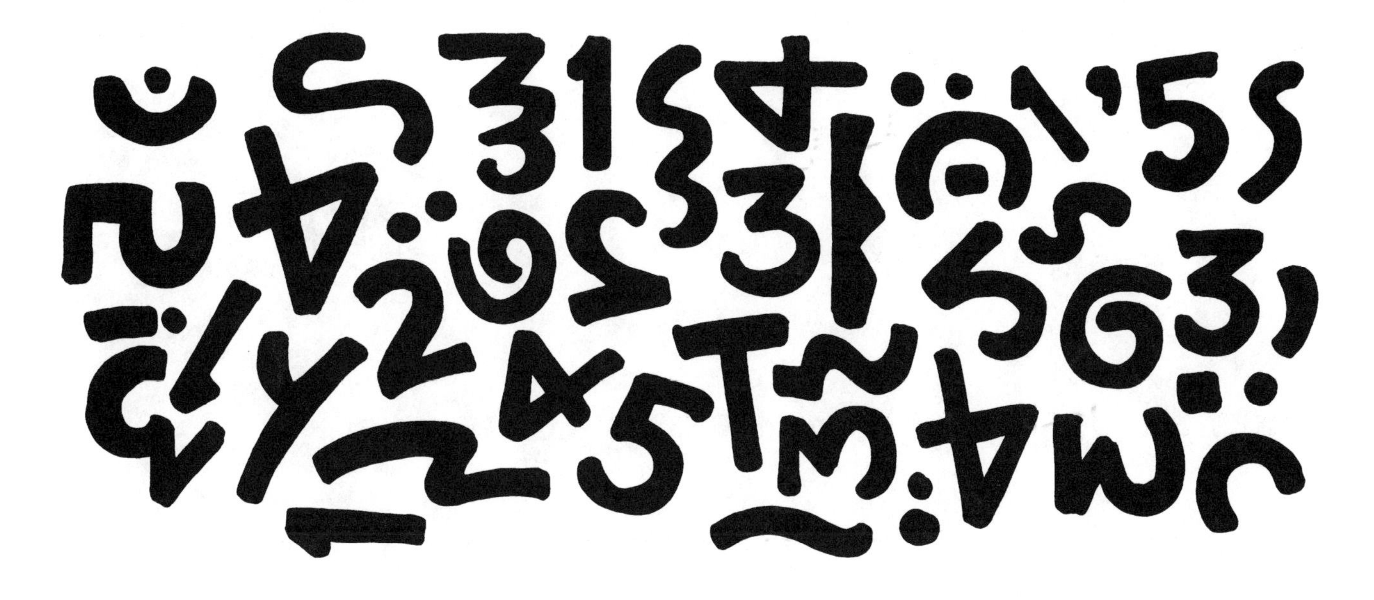

There are ____________ **1**s.

There are ____________ **2**s.

There are ____________ **3**s.

There are ____________ **4**s.

There are ____________ **5**s.

Look at the numerals on this page. They are all mixed up! Some of them are right-side up, some are upside down, and some are even sideways! How many of each numeral can you find? Write your answers in the spaces below.

There are ____________ **6**s.

There are ____________ **7**s.

There are ____________ **8**s.

There are ____________ **9**s.

Hint:
The number **6** has a rounded top and the number **9** has a long, straight stem.

One number looks the same upside down as right-side up. Which number? ____________________

Each clown has three numbered beanbags. The numbers on each clown's set of beanbags are in **sequence,** or number order. Fill in the missing number in each set of beanbags.

When you play dominoes, you must make sure that the two sides of the dominoes that touch each other have the same number of dots. Fill in the missing dots on the dominoes on these two pages.

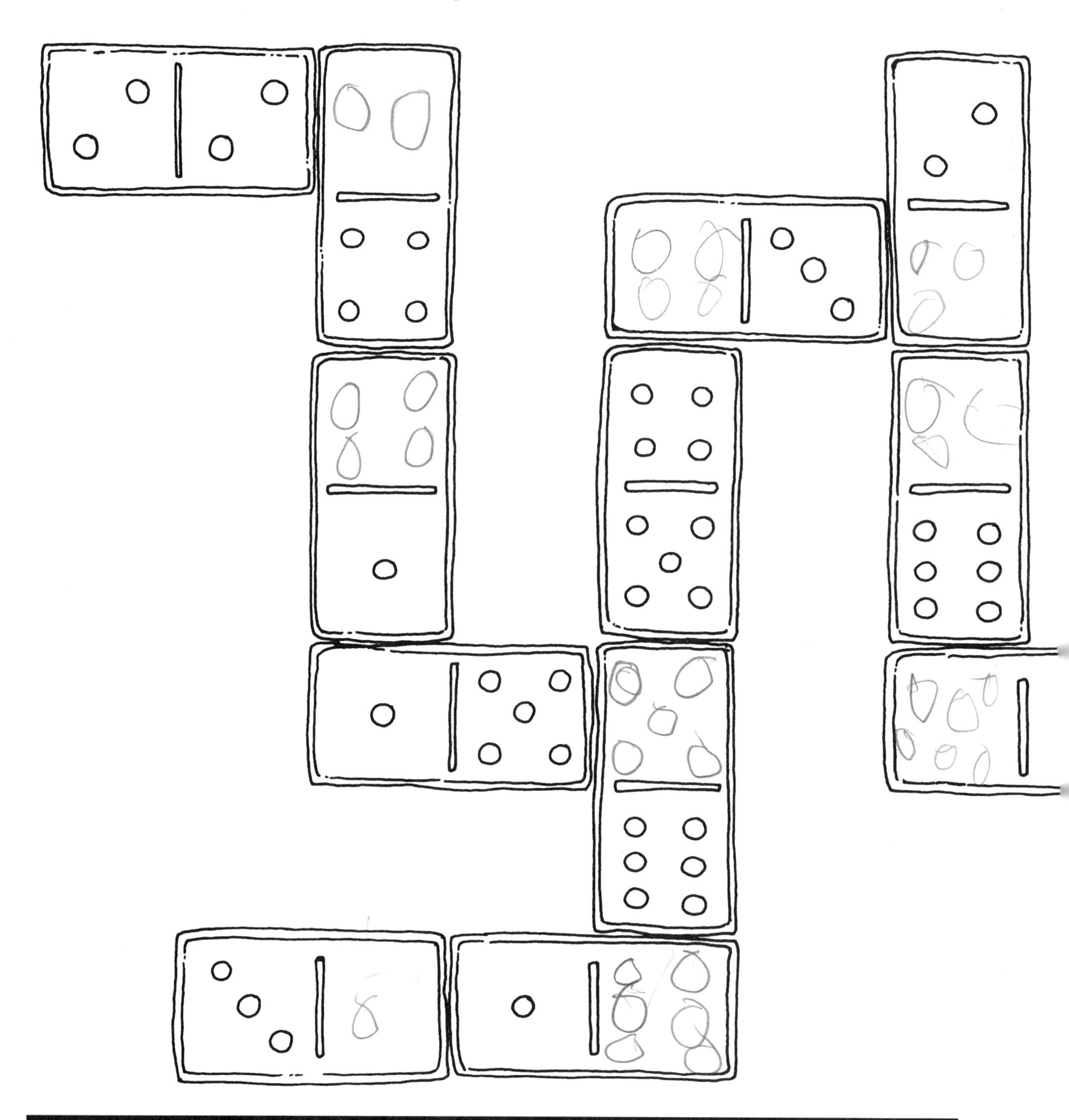

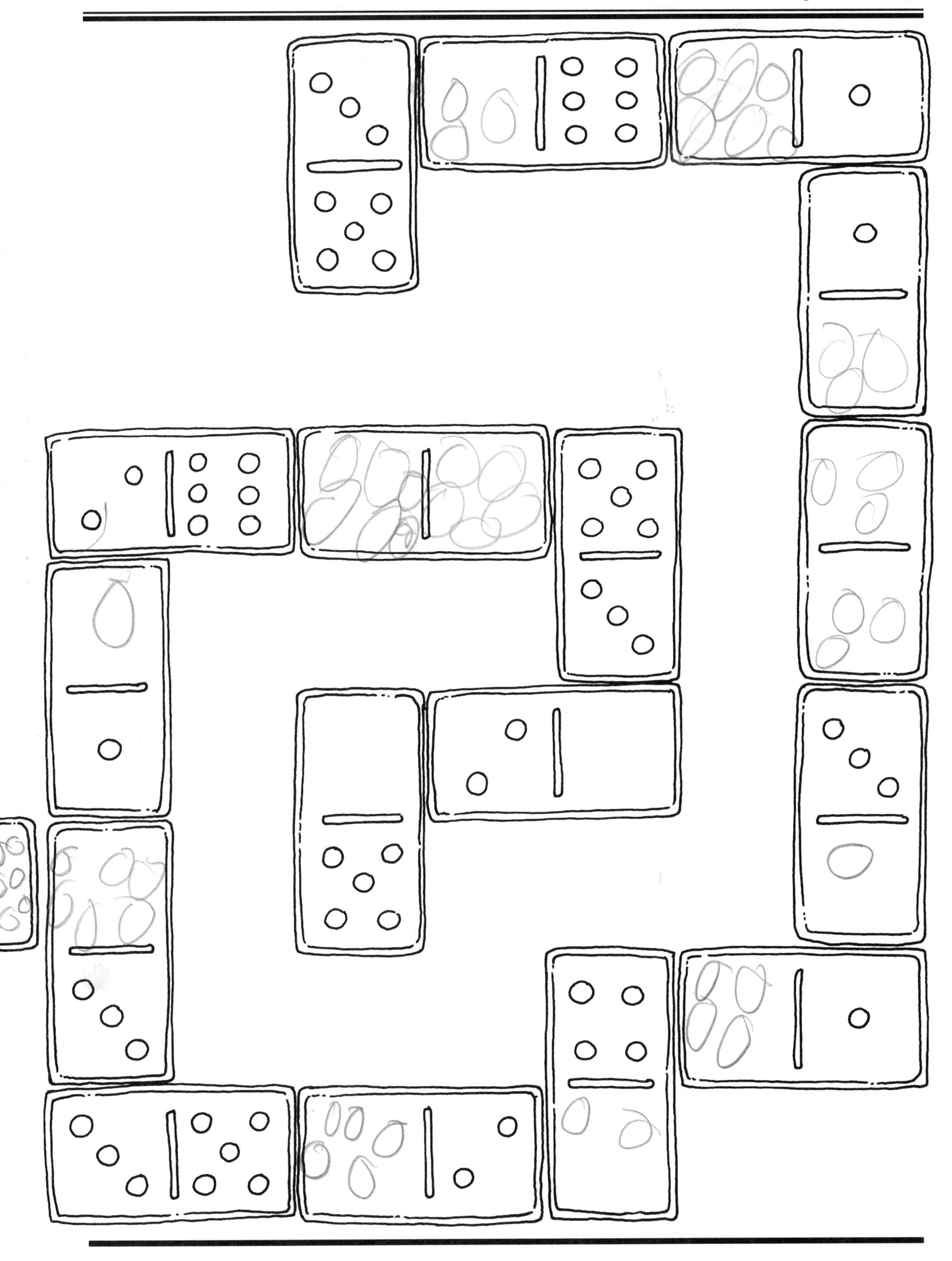

Use the following numbers to fill in the shapes below so that each statement is true. Use each number only once.

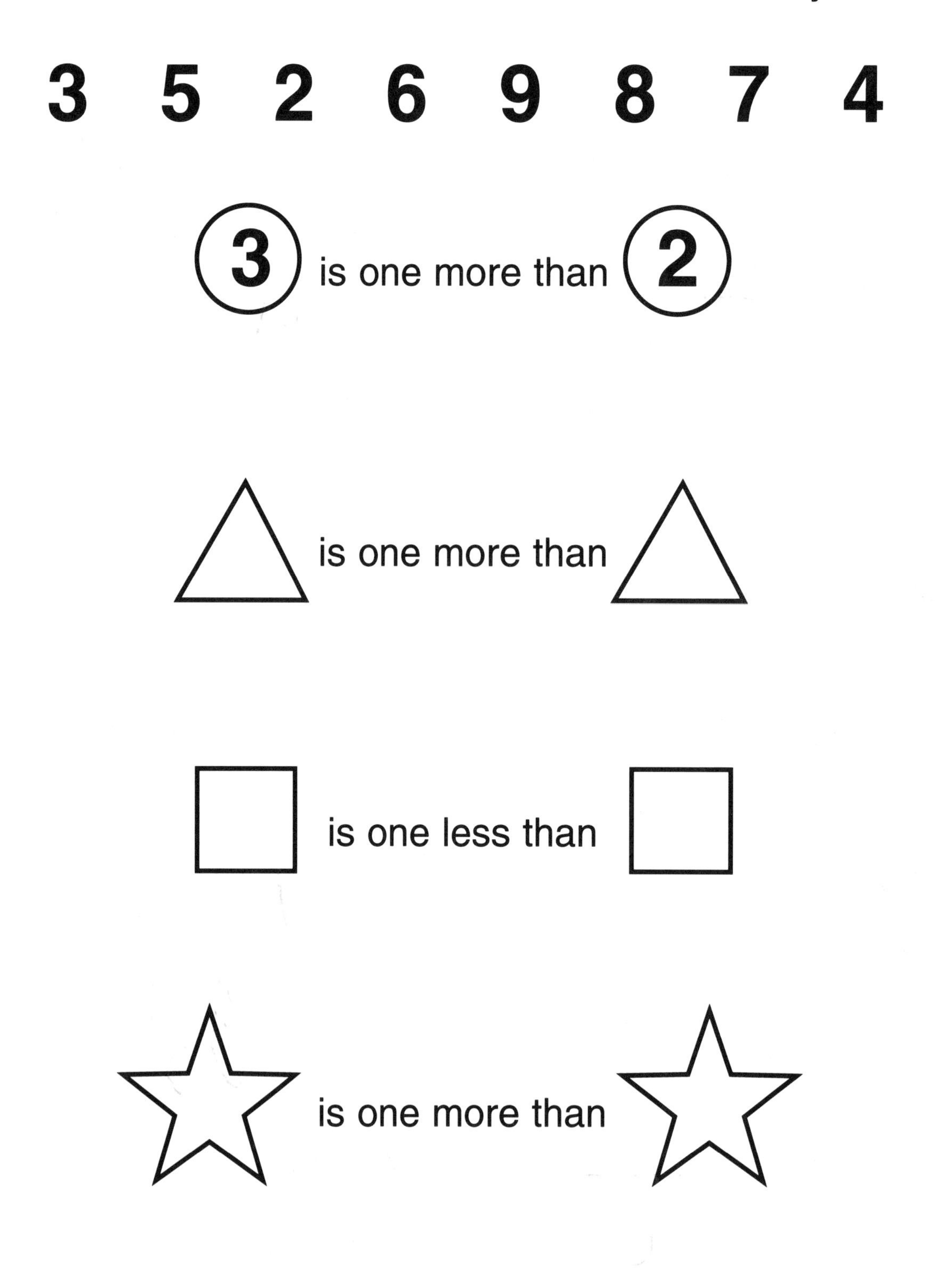

Fill in the blanks in the sentences below with the word **more** or the word **less**. Use the pictures to help you.

There are __________ girls than boys on the team.

The team has __________ balls than bats.

There are __________ people standing than sitting.

__________ people are wearing hats than sunglasses.

Fill in the blanks in the sentences below with the word **more** or the word **less**. Use the pictures to help you.

There are __________ seals than penguins playing in the snow.

There are __________ big penguins than little penguins.

__________ polar bears are swimming than resting.

There are __________ fish than polar bears swimming.

Find the secret number using the clues below. Put an **X** on each number that you **eliminate,** or do not think is the secret number.

Hint: Even numbers can be divided in half equally. Odd numbers cannot be divided in half equally.

Even number:

Odd number:

CLUES

- The number is greater than 4.
- The number is even.
- The number is less than 7.

What is the secret number? __________

Find the secret number using the clues below. Put an **X** on each number that you **eliminate,** or do not think is the secret number.

Hint: Even numbers can be divided in half equally. Odd numbers cannot be divided in half equally.

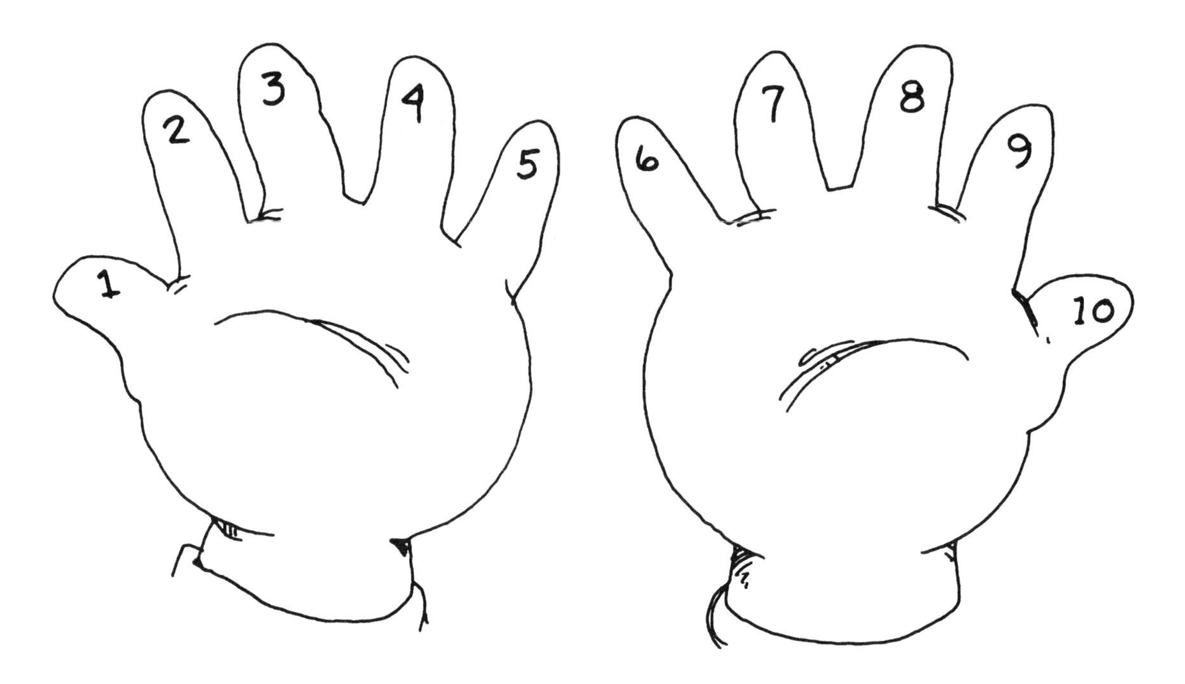

CLUES

- The number does not include the numeral 1.
- The number is not even.
- The number is less than the number of fingers on one hand.
- The number is greater than the number of ears a person has.

What is the secret number? __________

Find the teddy bear who is wearing the secret number using the clues below. Put an **X** on each teddy bear that you **eliminate,** or do not think is wearing the secret number.

Hint: Even numbers can be divided in half equally. Odd numbers cannot be divided in half equally.

CLUES

- The teddy bear with the secret number is not at the beginning of any row.
- The teddy bear with the secret number has an odd number on his tummy.
- The teddy bear with the secret number is not in the first row.
- The teddy bear with the secret number is at the end of a row.

What is the secret number? __________

Find the egg with the secret number using the clues below. Put an **X** on each egg that you **eliminate,** or do not think has the secret number.

Hint: Even numbers can be divided in half equally. Odd numbers cannot be divided in half equally.

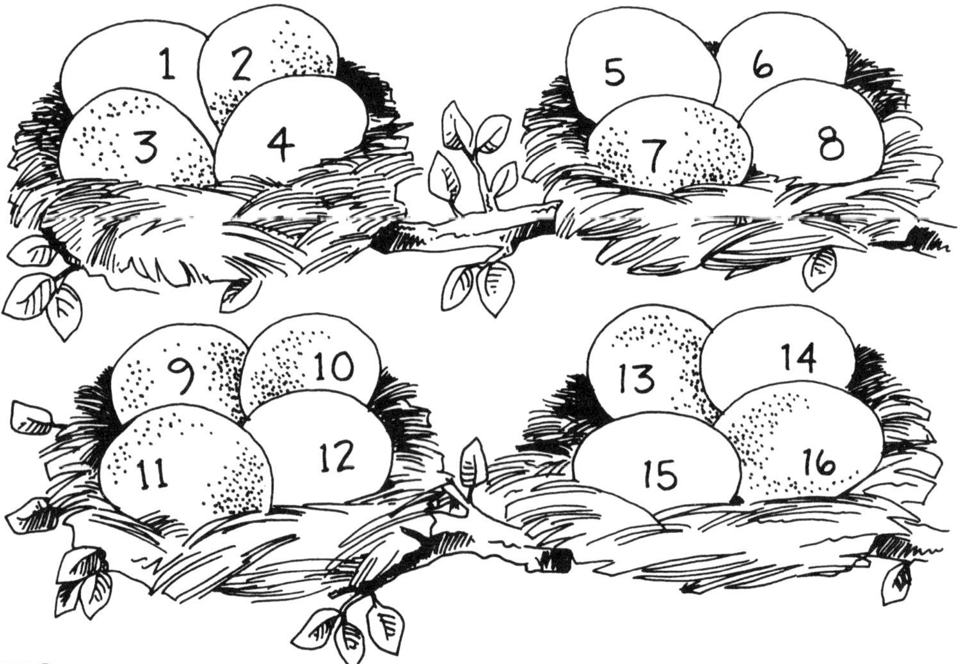

CLUES

- The egg with the secret number is not in the second nest.
- The egg with the secret number is not speckled.
- The egg with the secret number has an even number.
- The egg with the secret number is not in the last nest.
- The egg with the secret number has a number that is higher than 5.

What is the secret number? __________

Find the apple with the secret number using the clues below. Put an **X** on each apple that you **eliminate,** or do not think has the secret number.

CLUES

- The apple with the secret number is not on the shortest tree.
- The apple with the secret number is not one of the largest apples.
- The apple with the secret number is smallest on its tree.
- The apple with the secret number is not on the tallest tree.

What is the secret number? __________

For this game you will need a small cup full of unpopped popcorn kernels.

Look at the square below. Guess how many popcorn kernels it will take to fill in the square. This kind of guess is called an **estimate**. Write your estimate here: __________

Now fill the square with the popcorn kernels. Try not to leave any white space showing between the kernels.

Count how many kernels you used. Write that number here. __________

Was your estimate, or guess, more, less, or exactly the same as the actual number of kernels that fit in the square? ____________________________

For this game you will need a small cup full of unpopped popcorn kernels.

Look at the square below. It is bigger than the square on the previous page. Guess how many popcorn kernels it will take to fill in the square. This kind of guess is called an **estimate**. Write your estimate here: __________

Now fill in the square with the kernels. Try not to leave any white space showing between the kernels.

Count how many kernels you used. Write that number here. __________

Was your estimate, or guess, more, less, or exactly the same as the actual number of kernels that fit in the square? ____________________________

Fergie Frog wants to go for a hop along the riverbank. Use the number line to figure out all the places he stops.

Fergie takes 3 hops forward and 2 more hops forward. Where does he land? ______________________

After resting Fergie takes 1 hop forward and 4 hops backward. Where does he land? ______________________

He then takes 2 hops forward and 1 hop backward. Where does he land? ______________________

Next Fergie takes 5 hops forward and 1 more hop forward. Where does he land? ______________________

Then Fergie takes 2 hops backward. Where does he land?

____

Finally Fergie takes 1 hop forward and 8 hops backward.
Where does he land? ____________________________

Now tell a story to a friend or family member about Fergie and his friends. How do they spend their days?

Use orange and green crayons to color the gummy worms in the box below. Make sure that you have an equal number of green and orange gummy worms.

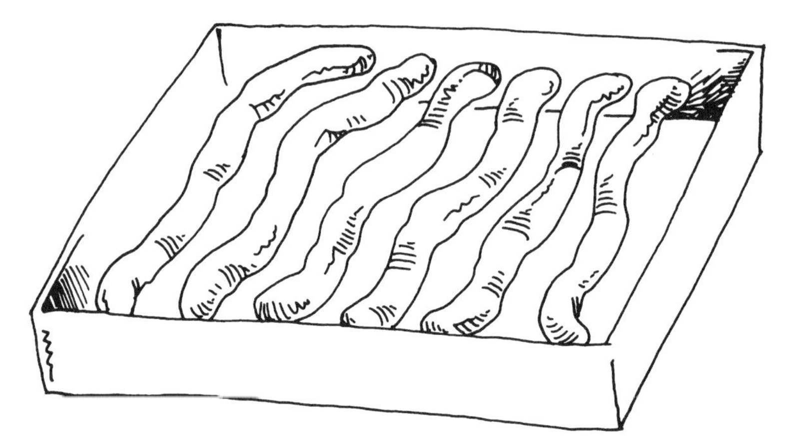

If you gave all of the green gummy worms to a friend, how many gummy worms would be left for you? ___________

Use red and yellow crayons to color the cherries in the box below. Make sure that you have an equal number of red and yellow cherries.

If you ate all of the red cherries, how many cherries would you have left? ___________

Use pink and purple crayons to color the cupcakes. Make sure that you have an equal number of pink and purple cupcakes.

If you gave 2 of the pink cupcakes and 3 of the purple cupcakes to your friends, how many cupcakes would be left? ____________

How many pink cupcakes would be left? ____________

How many purple cupcakes would be left? ____________

Ask an adult to help you find a small cup and 5 large, dried lima beans. Use a marker to color one side of each bean red.

1. Place the beans in the cup.
2. Gently shake the cup and then pour the beans out on the table.
3. Count the beans that landed with the red side facing up. Count the beans that landed with the white side facing up.
4. Record your results by coloring in the first row of beans below.
5. Repeat this experiment seven more times. Each time record your results below.

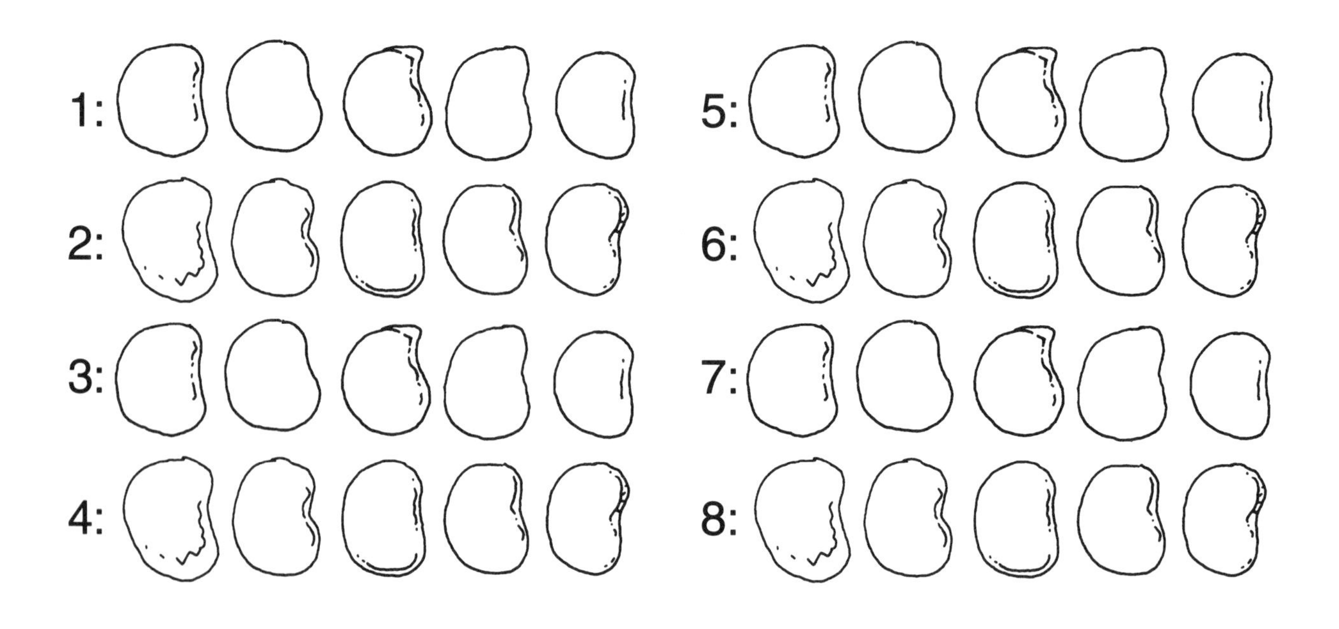

Now look at your results. How many times did each of the following combinations of beans occur?

Five of one color and none of the other color: __________

Four of one color and one of the other color: __________

Three of one color and two of the other color: __________

If you played this game again, do you think your results would be exactly the same, about the same, or extremely different? Circle your answer.

Why do you think so? ______________________________

____

____

____

Play again and test your prediction!

Draw 1 bird in the tree. Then draw another bird in the tree. Use crayons to color each bird a different color.

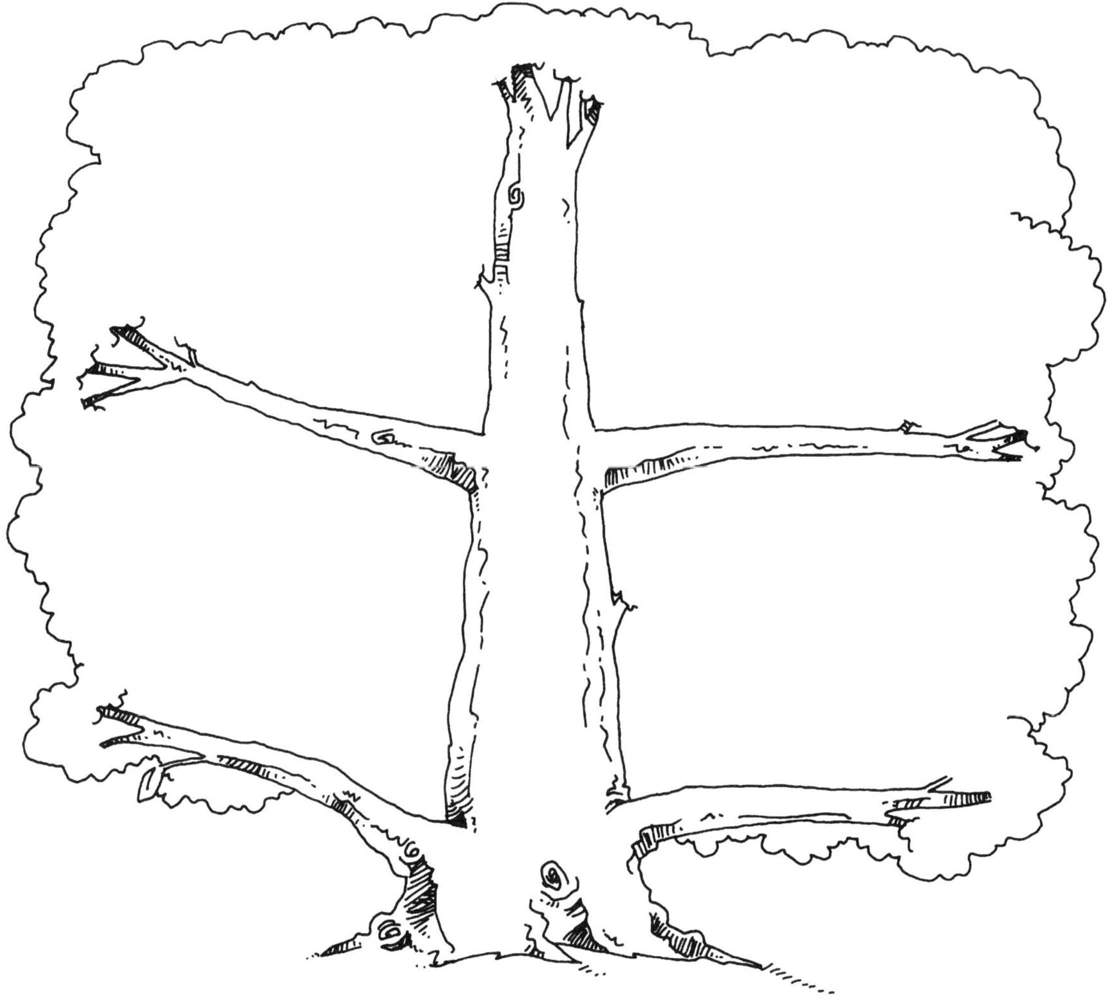

Use the clues in the picture to help you complete the following number sentences.

1 bird + 1 bird = ____________________

1 + 1 = ___________

Now tell a story about the picture. Be sure to include some numbers in your story!

A **pair** is a group of two. Draw one pair of fish in the aquarium below. Then draw another pair of fish. Use crayons to color each pair a different color.

Use the clues in the picture to help you complete the following number sentences.

2 fish + 2 fish = ________________

2 + 2 = ______

Now tell a story about the picture. Be sure to include some numbers in your story!

Draw one group of 3 raindrops falling from the cloud below. Then draw another group of 3 raindrops. Use crayons to color each group of raindrops a different color.

Use the clues in the picture to help you complete the following number sentences.

3 raindrops + 3 raindrops = ______________________

3 + 3 = ________

Now tell a story about the picture. Be sure to include some numbers in your story!

Draw one group of 4 ants on the anthill below. Then draw another group of 4 ants. Use crayons to color each group of ants a different color.

Use the clues in the picture to help you complete the following number sentences.

4 ants + 4 ants = ____________________

4 + 4 = ________

Now tell a story about the picture. Be sure to include some numbers in your story!

Draw one group of 5 bees on the beehive. Then draw another group of 5 bees. Use crayons to color each group of bees a different color.

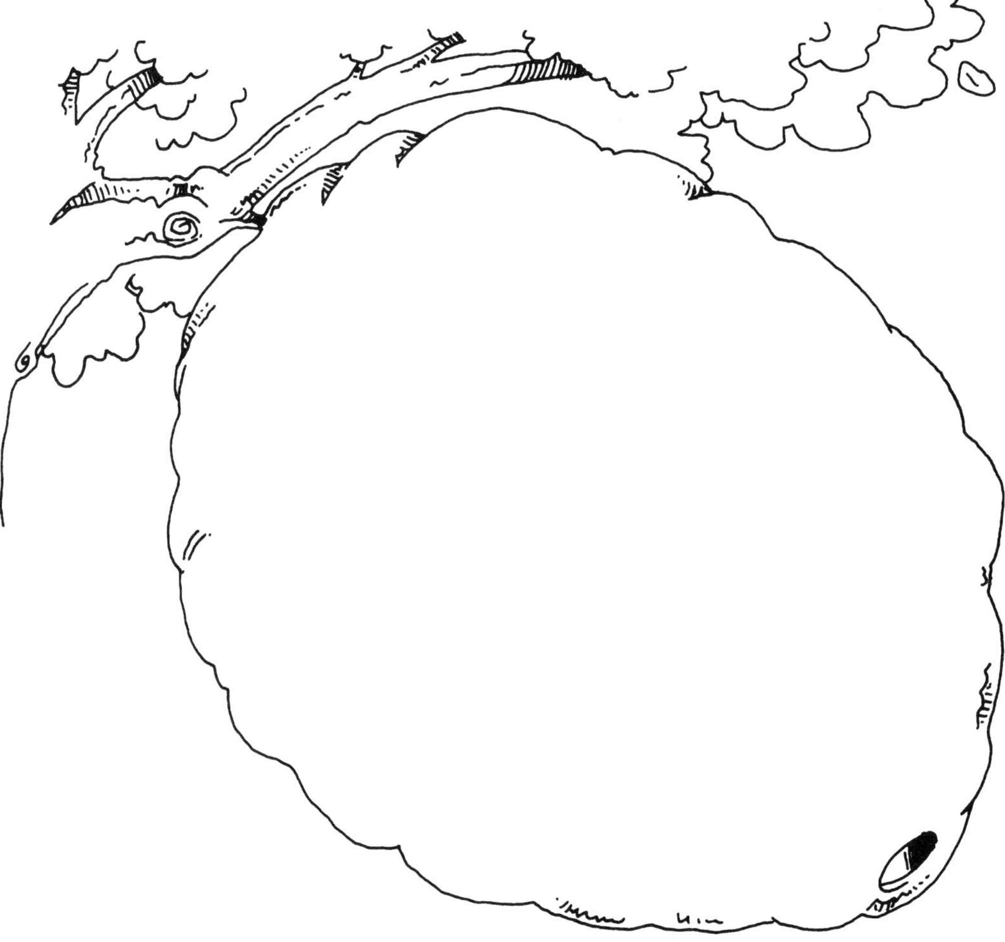

Use the clues in the picture to help you complete the following number sentences.

5 bees + 5 bees = ______________________

5 + 5 = ________

Now tell a story about the picture. Be sure to include some numbers in your story!

Draw a line from each number sentence to the picture which shows what the number sentence says.

1 + 2 = 3

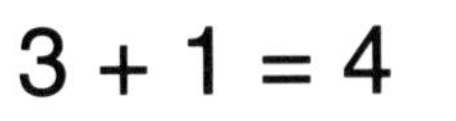

3 + 1 = 4

2 + 2 = 4

3 + 0 = 3

A **number chain** is a series of two or more numerals. Find and circle each number chain described below. Number chains can run from left to right or from top to bottom. The first one is done for you.

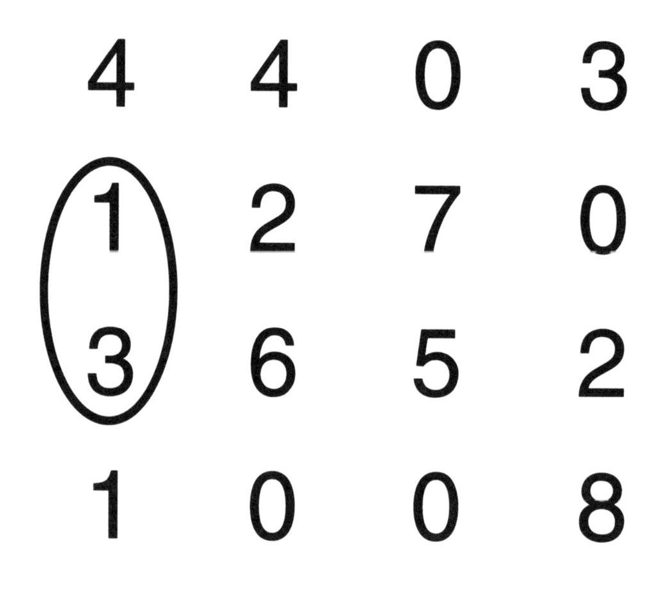

The number that is 1 more than 12.

The number of days in a year.

The number of months in a year.

The number that equals three 10s.

The number of pennies
in one dollar.

A **number chain** is a series of two or more numerals. Find and circle each number chain described below. Number chains can run from left to right or from top to bottom.

6	7	1	8
0	3	9	2
5	0	7	4
1	3	2	0

The number that is 1 more than 19.

The number that is 1 less than 25.

The number that is more than 38 but less than 40.

The number of seconds in one minute.

The number of states in the United States.

A **number chain** is a series of two or more numerals. Find and circle each number chain described below. Number chains can run from left to right or from top to bottom.

Hint: Use a calendar to help you.

2 2 0 3

6 8 4 6

3 0 2 5

3 1 4 9

The number that equals the number of cents in a quarter.

The number that is greater than 13 but less than 15.

The number that is 2 more than 20.

The number that tells how many days are in the month of June.

The number that tells how many days are in the month of January.

Ask an adult to help you find 6 buttons, pennies, beans, or other small objects around the house. Divide the objects into two groups. Place 2 objects in one group and 4 in another group. They should look like this:

Now find three more ways to group the buttons or small objects. In the boxes below, draw pictures of the three new ways that you grouped your objects.

Use the picture to help you answer the questions below.

Sally bought a pair of roller skates. How many wheels did she buy? ______ Explain your answer.________________

____

____

____

Dee bought 3 bicycles and 1 skateboard. How many wheels did she buy? ______ Explain your answer.______________

____

____

____

Use the pictures to help you answer the questions below.

Farmer Ralph has cows and chickens. Together his animals have a total of 12 legs and 4 horns. How many cows does Farmer Ralph have? ______ How many chickens does he have? ______ Explain your answer. ________________

____

____

____

Farmer Ben also has cows and chickens. Together his animals have a total of 16 legs and 6 horns. How many cows does Farmer Ben have? ______ How many chickens does he have? ______ Explain your answer. ___________

____

____

____

Trace the worms below onto another sheet of paper. Use crayons to color the worm named **Greenie** green and the worm named **Pinky** pink.

Use your worms to measure the objects on these two pages. Record your answers in the spaces provided.

How many **Greenie** lengths long is the log? ___________

How many **Pinky** lengths long is the log? ___________

How many **Greenie** lengths long is the garden hose that runs across both pages? ________

How many **Pinky** lengths long is the garden hose that runs across both pages? ________

How many **Greenie** lengths tall is the ladder? ___________

How many **Pinky** lengths tall is the ladder? ____________

How many **Greenie** lengths tall is the tree? ____________

How many **Pinky** lengths tall is the tree? ______________

How many **Greenie** lengths tall is the sunflower? _______

Without measuring, write down how many **Pinky** lengths tall you think the sunflower is. ____________

How can you tell without measuring? _________________

____

These two little mice, Moe and Molly, are very good friends. They always share the goodies they find equally. Use crayons to draw a blue circle around Moe's portion of goodies in each picture. Draw a red circle around Molly's portion of goodies in each picture.

Now make up a story about what would happen if Moe and Molly found 10 peanuts.

The girls are filling the bird feeder with birdseed. Use crayons to finish the picture and solve the problem below.

Ramona puts in 2 cups of sunflower seeds.

Kiki puts in 1 cup of millet.

Susan puts in 1 cup of cracked corn.

How much birdseed do they use altogether? ___________

Do the girls have enough seed to fill up the feeder completely? ___________

How many more cups of seed do they need? ___________

The Bateman children want to combine their money to buy a pet. Help them figure out which pets they can buy.

Hint: Draw a picture of the bills and coins each child has. Then count how much money they have altogether.

Camille has 3 dollar bills.

Danielle has 1 dollar bill and 4 quarters.

Andy has 2 dollar bills.

Which pets could the children buy with the money they have? ______________________________

Color those pets.

Which pet do you think they should buy? ____________

Why? ______________________________

Jean went to the library at 6:00. She stayed for 2 hours.

Draw a picture of Jean's watch showing the time when she left the library. Look at a real watch or clock if you need help.

Tom finished his workout at 8:00. His workout took 1 hour to complete.

Draw a picture of the clock on the gym wall showing the time when he began his workout.

Gary began hiking at 8:00. He arrived at the top of the mountain in 3 hours.

Draw a picture of a clock in the box showing the time when Gary reached the top of the mountain. Look at a real watch or clock if you need help.

Peggy planted 5 rosebushes. It took her a ½ hour to plant each rosebush. She started planting at 9:00 and did not take a break until **all** of the rosebushes were planted.

Draw a picture of a clock in the box showing the time when Peggy finished.

Answers

Page 5

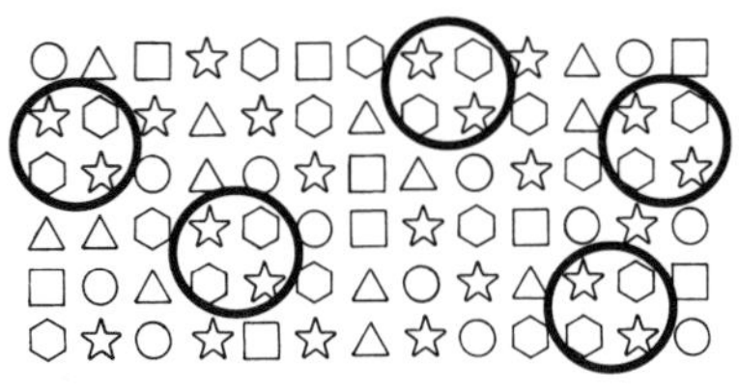

5 clusters

Page 6

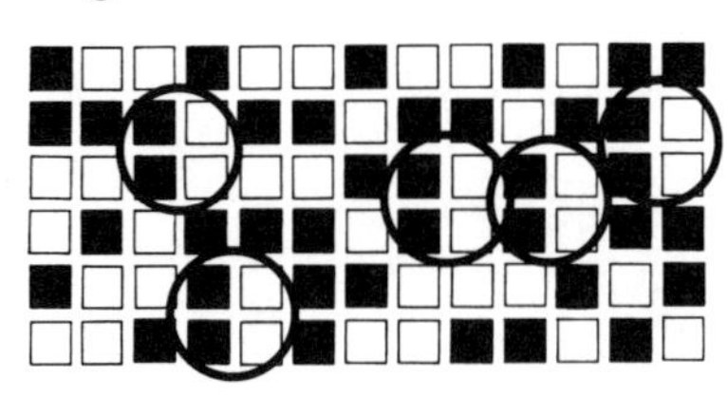

5 clusters

Page 7

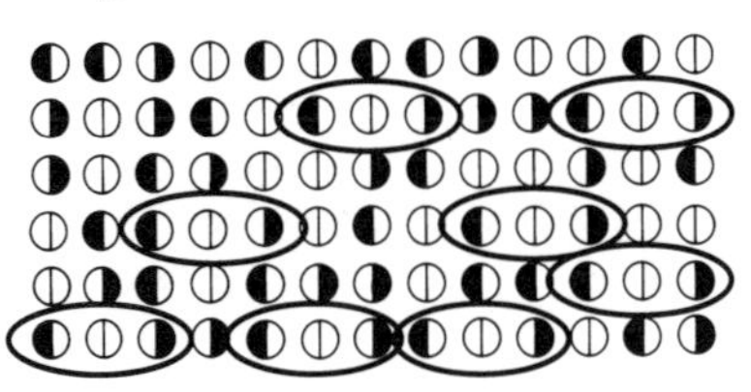

8 clusters
Rest of answer will vary.

Page 8

Page 9

Page 10

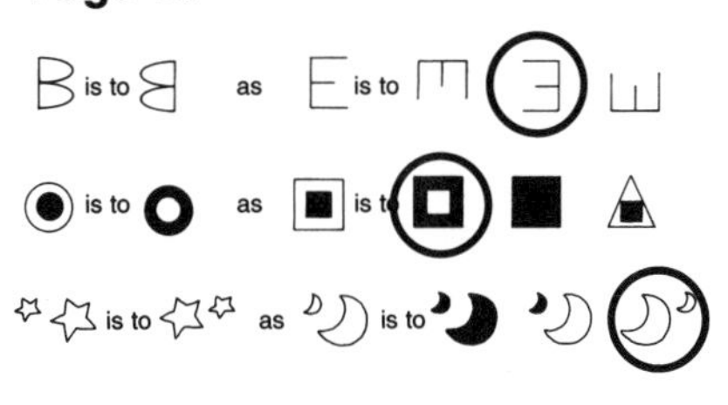

Page 11

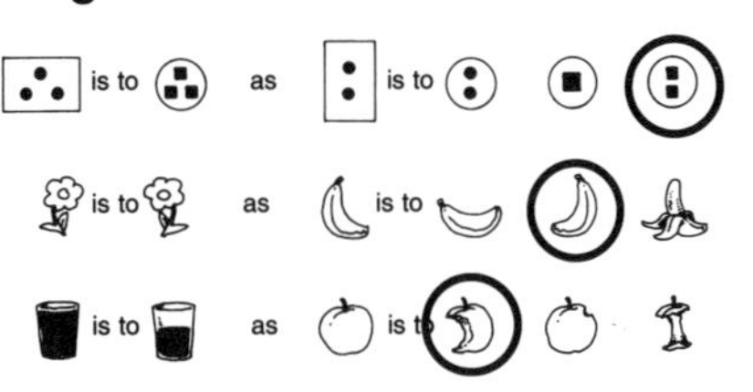

Pages 12–13

Answers will vary. Pattern on socks should be continued across both pages. Each line of laundry should be colored to create a repeating pattern.

Page 14

Child is most likely to pick a chocolate cookie because there are about one-third more chocolate cookies than vanilla ones.

Page 15

Answers will vary, but may include:

blue / green / blue / blue

blue / green / blue / green

green / blue / green / green

blue / green / green / blue

Page 16

Answers will vary, but may include:

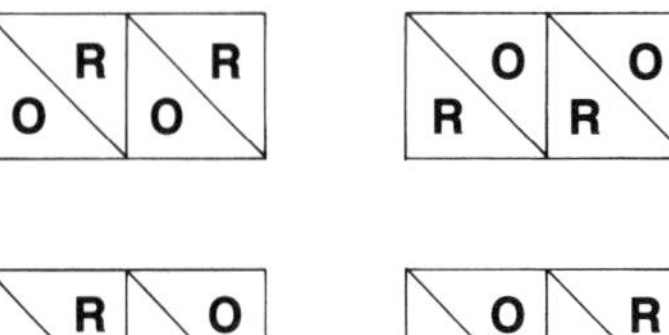

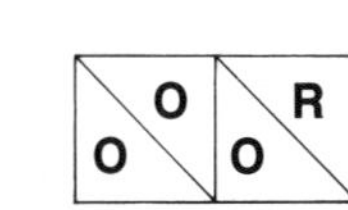

R R O R / O O O O

Page 17

2

3

4

4

10

Page 18

Answers will vary, but may include the following observations:
Both houses have one door and windows. Each has a triangular roof and one chimney.

The houses have a different number of windows. Some of the windows are different shapes. The doors are on opposite sides of the houses and have different shaped doorknobs. Michael's house has bushes on either side of it and the other house has trees on each of its sides.

Page 19
Answers will vary, but may include the following observations: The cereal box, bread loaf, and butter are rectangular in shape. Each has six sides (also called faces) and four corners. The drinking glass and pitcher each hold liquid. The glass and salt and pepper shakers are cylindrical in shape.

The objects differ in size. The box is hollow, but the other objects are solid. The objects vary in the purposes for which they are used, as well as in the materials from which they are made.

Page 20
Answers will vary, but may include the following observations: The ice-cream cones are both cold and sweet. They are both the same color. Both of the cones have lines on them.

The cones are different shapes. One has two scoops and one is soft serve. One cone has lines running in two directions, forming a grid. The other cone has diagonal lines on it.

Page 21
4, 1, 4, 3, 2

Page 22
3, 4, 8, 4
The numeral 8 looks the same right-side up and upside down.

Page 23
2, 4, 9

Pages 24–25

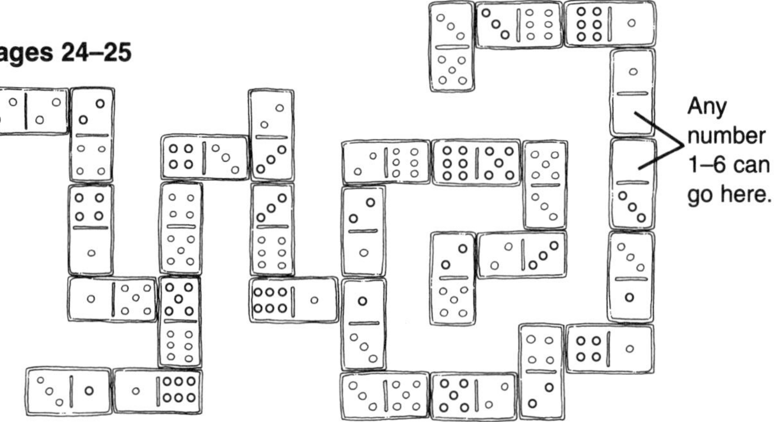

Page 26
5 is one more than 4
6 is one less than 7
9 is one more than 8

Page 27
more
more

less
less

Page 28
less
more

more
less

Page 29
6

Page 30
3

Page 31
15

Page 32
12

Page 33
6

Page 34
Answers will vary.

Page 35
Answers will vary.

Pages 36–37
5 (the pine tree)
2 (the ladybug's house)
3 (the beaver dam)
9 (the anthill)
7 (the spider web)
0 (back at home)

Page 38
3 gummy worms
4 cherries

Page 39
5 cupcakes
3 pink
2 purple

Pages 40–41
Answers will vary.

Page 42
1 bird + 1 bird = 2 birds
1 + 1 = 2

Page 43
2 fish + 2 fish = 4 fish
2 + 2 = 4

Page 44
3 raindrops + 3 raindrops = 6 raindrops
3 + 3 = 6

Page 45
4 ants + 4 ants = 8 ants
4 + 4 = 8

Page 46
5 bees + 5 bees = 10 bees
5 + 5 = 10

Page 47

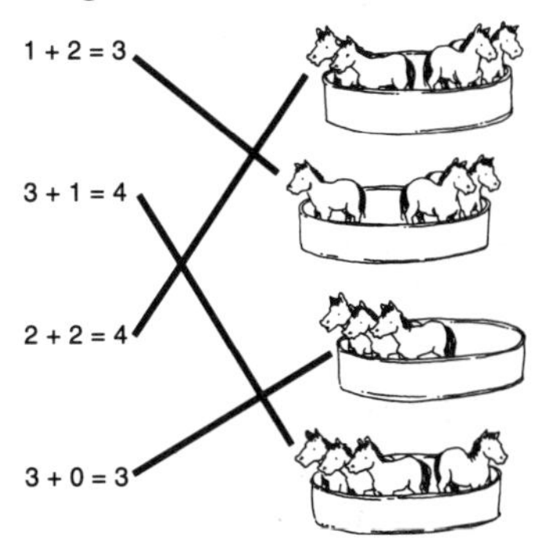

Page 48

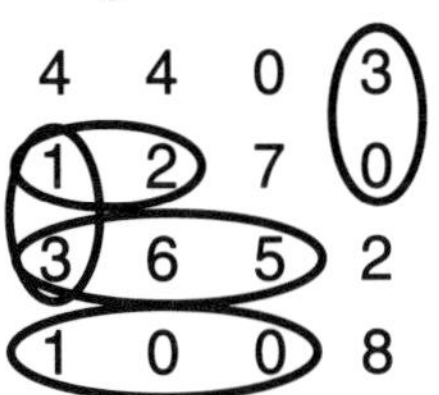

13 is one more than 12.
There are 365 days in a year.
There are 12 months in a year.
30 equals three 10s.
There are 100 pennies in one dollar.

Page 49

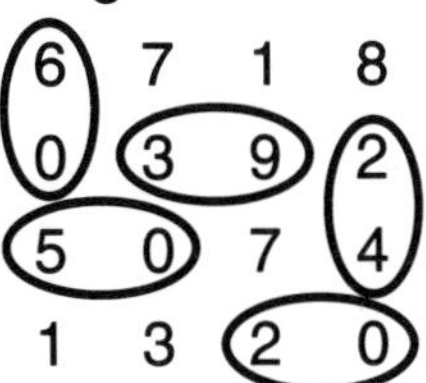

20 is 1 more than 19.
24 is 1 less than 25.
39 is more than 38 but less than 40.
There are 60 seconds in one minute.
There are 50 states in the United States.

Page 50

2 2 0 3
6 8 4 6
3 0 2 5
3 1 4 9

There are 25 cents in a quarter.
14 is greater than 13 but less than 15.
22 is 2 more than 20.
There are 30 days in the month of June.
There are 31 days in the month of January.

Page 51

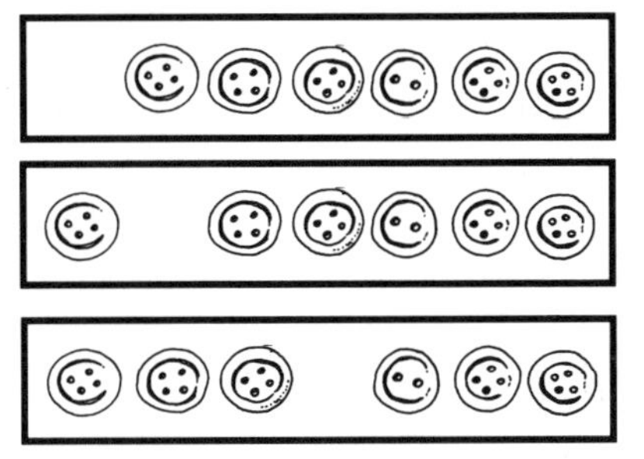

Page 52
Sally bought 2 skates. There are 4 wheels on each skate. Sally bought 8 wheels.

Dee bought 3 bicycles. There are 2 wheels on each bicycle. She bought 1 skateboard. There are 4 wheels on a skateboard. Dee bought 10 wheels.

Page 53
2 cows
2 chickens

3 cows
2 chickens

Pages 54–55
Log: 4 Greenies, 2 Pinkys
Garden hose: 10 Greenies, 5 Pinkys

Pages 54–55 (continued)
Ladder: 3 Greenies, 1½ Pinkys
Tree: 4 Greenies, 2 Pinkys
Sunflower: 2½ Greenies, 1¼ Pinkys

You can tell because Pinky is half as long as Greenie.
It takes 2 Greenie lengths to equal 1 Pinky length.

Page 56

Rest of answer will vary but should show that Moe and Molly each get 5 peanuts.

Page 57
4 cups. No. They need 1 more cup.

Page 58
The children have $7.00. They could buy the bird, the iguana, or the kitten.
Rest of answer will vary.

Page 59

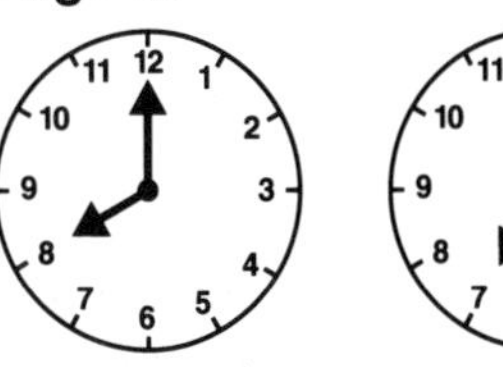

Page 60

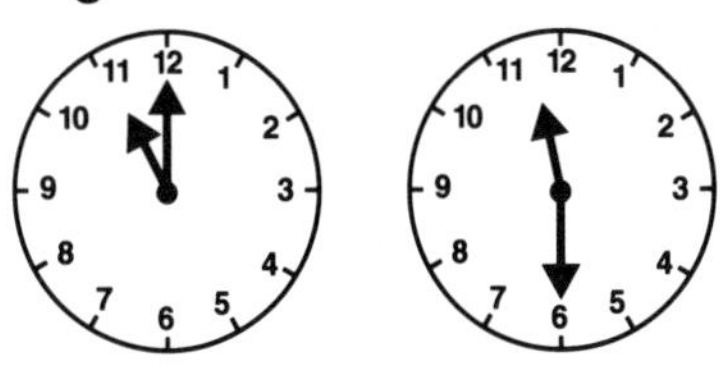